Psychology of Programming

Computers and People Series

Edited by

B. R. GAINES and A. MONK

Monographs

Communicating with microcomputers: An introduction to the technology of man–computer communications, *Ian H. Witten* 1980

The Computer in Experimental Psychology, *R. Bird* 1981

Principles of Computer Speech, *I. H. Witten* 1982

Cognitive Psychology of Planning, *J-M. Hoc* 1988

Edited Works

Computing Skills and the User Interface, *M. J. Coombs and J. L Alty (eds)* 1981

Fuzzy Reasoning and Its Applications, *E. H. Mamdani and B. R. Gaines (eds)* 1981

Intelligent Tutoring Systems, *D. Sleeman and J. S. Brown (eds)* 1982 (1986 paperback)

Designing for Human–Computer Communications, *M. E. Sime and M. J. Coombs (eds)* 1983

The Psychology of Computer Use, *T. R. G. Green, S. J. Payne and G. C. van der Veer (eds)* 1983

Fundamentals of Human–Computer Interaction, *A. Monk (ed)* 1984, 1985

Working with Computers: Theory versus Outcome, *G. C. van der Veer, T. R. G. Green, J-M. Hoc and D. Murray (eds)* 1988

Cognitive Engineering in Complex Dynamic Worlds, *E. Hollnagel, G. Mancini and D. D. Woods (eds)* 1988

Computers and Conversation, *P. Luff, N. Gilbert and D. Frohlich (eds)* 1990

Adaptive User Interfaces, *D. Browne, P. Totterdell and M. Norman (eds)* 1990

EACE Publications

(Consulting Editors: *Y. WAERN and J-M. HOC)*

Cognitive Ergonomics, *P. Falzon (ed)* 1990

Psychology of Programming, *J-M. Hoc, T. R. G. Green, R. Samurçay and D. J. Gilmore (eds)* 1990

Psychology of Programming

edited by

J.-M. Hoc
CNRS - Université de Paris 8, Psychologie Cognitive du Traitement de l'Information Symbolique, Paris, France

T. R. G. Green
MRC Applied Psychology Unit, Cambridge, UK

R. Samurçay
Université de Paris 8, Psychologie Cognitive du Traitement de l'Information Symbolique, Paris, France

D. J. Gilmore
Department of Psychology, University of Nottingham, Nottingham, UK

A Publication of the
European Association of Cognitive Ergonomics

ACADEMIC PRESS
Harcourt Brace Jovanovich, Publishers
London San Diego New York
Boston Sydney Tokyo Toronto

ACADEMIC PRESS LTD.
24-28 Oval Road,
London NW1 7DX

United States edition published by
ACADEMIC PRESS INC.
San Diego, California 92101-4311

This book is printed on acid-free paper

British Library Cataloguing in Publication Data

is available

ISBN 0-12-350772-3

Acknowledgement

This book has been edited with the financial support of the PIRTTEM of CNRS and the association "Naturalia et Biologia".

Typeset by Keywords, Glasgow
Printed in Great Britain by St Edmundsbury Press Ltd,
Bury St Edmunds, Suffolk

Contents

Contributors

P. Brna
Department of Artificial Intelligence, University of Edinburgh, UK

R. Carn
Reliability Consultants Limited, Fearnside, Little Park Farm Road, Segensworth, Fareham PO15 5SH, UK

B. Curtis
Microelectronics and Computer Technology Corporation, 9430 Research Boulevard, Austin, Texas 78759, USA

F. Détienne
Projet de Psychologie, Ergonomique pour L'Informatique INRIA, Rocquencourt, BP 105, 78153, Le Chesnay Cedex, France

D. J. Gilmore
Psychology Department, University of Nottingham, Nottingham NG7 2RD, UK

B. Grabowski
Department of Anthropology, Rawles Hall, Indiana University, Bloomington, Indiana 47402, USA

T. R. G. Green
MRC Applied Psychology Unit, 15 Chaucer Road, Cambridge, CB2 2EF, UK

J.-M. Hoc
CNRS-Université de Paris 8 URA 1297, Psychologie Cognitive du Traitement de L'Information Symbolique, 2 Rue de la Liberté, F-93526 Saint Denis Cedex 2, France

B. Kitchenham
National Computing Centre, Oxford Road, Manchester, UK

P. Mendelsohn
Faculty of Psychology and Educational Science, University of Geneva, Switzerland

A. Nguyen-Xuan
CNRS-Université de Paris 8 URA 1297, Psychologie Cognitive du Traitement de L'Information Symbolique, 2 Rue de la Liberté, F-93526 Saint Denis Cedex 2, France

T. Ormerod
Department of Human Sciences, Loughborough University of Technology, Loughborough, Leicestershire, LE11 3TU, UK

C. Pair
Centre de Recherche en Informatique de Nancy, BP 239, 54506 Vandœuvre, France

N. Pennington
Psychology Department, Campus Box 345, University of Colorado, Boulder, Colorado 80309, USA

M. Petre
Instituut voor Perceptie Onderzoek/IPO, Postbus 513, 5600 MB Eindhoven, The Netherlands

J. Rogalski
CNRS-Université de Paris 8 URA 1297, Psychologie Cognitive du Traitement de L'Information Symbolique, 2 Rue de la Liberté, F-93526 Saint Denis Cedex 2, France

R. Samurçay
CNRS-Université de Paris 8 URA 1297, Psychologie Cognitive du Traitement de L'Information Symbolique, 2 Rue de la Liberté, F-93526 Saint Denis Cedex 2, France

W. Visser
INRIA-Ergonomics Psychology Group, Rocquencourt BP 105, F-78153, Le Chesnay, France

Diane Walz
Department of Accounting and Information Systems, University of Texas at San Antonio, San Antonio, TX 78285, USA

Part 1

Theoretical and Methodological Issues

Programming is a human activity that is a great challenge, involving the design of machine behaviour that can assist, and at times replace, humans in intellectual tasks. Seen in this light, programming is a meta-activity, and the study of programming is useful to the development of basic knowledge in cognitive psychology. For the most part, research works in the psychology of problem solving deal with result-production situations where subjects are asked to attain a goal without being required to express a general procedure to reach this state as is the case of program-production situations, such as sequencing in manufactures, designing directions for use, etc. (Miller, 1981; Hoc, 1988; Chapter 2.3). Computer programming is a valuable and rich paradigm for studying this latter type of situation. In turn, computer science can draw on psychological studies of programming to design more appropriate programming languages and environments as well as more efficient training curricula. For example, in a context of development of top-down programming methodologies, the observation that even expert programmers are not able to conform themselves to this kind of strategy, but show 'opportunistic' strategies which mix top-down and bottom-up components, may lead computer scientists to design more efficient support to programming which does not ignore this fact.

The history of the psychology of programming dates back to the 1960s, probably to a French work which analysed business programmers' behaviour in terms of top-down, data-oriented programming methodology (Rouanet and Gateau, 1968). The broad range of individual differences in the structuring of programs designed to solve the same problem led the authors to the conclusion that a programming methodology was needed for programmers to teach them to use high-level (more abstract) representations of information and control flow to correct for the saliency effects of low-level machine constraints. These results are consistent with more recent studies of planning which indicate that programmers need support to process schematic representations before implementing them in concrete devices (Hoc, 1988).

This kind of research was the exception rather than the rule in the early days of the psychology of programming. Most studies were produced by computer scientists who developed a normative approach to what they considered to be the most powerful programming concepts. The book by Dahl *et al.*(1972) on the principles of structured programming is a good example.

During the 1970s, the 'second generation' of programming studies was dominated by a number of empirical studies, comparing diverse types of languages (e.g. diverse expressions of conditionals), program layouts (e.g. flowchart, indentation, etc.), methods (e.g. top-down, modular, etc.), and so on. The watershed publication initiating this era was by Weinberg (1971) which argues that a psychological viewpoint should be incorporated into any approach to programming. The seminal changes in theoretical frameworks in the study of human problem solving had only just appeared (Newell and Simon, 1972), and Weinberg makes reference to aptitude theory which was better known at that time but was in fact ill suited to the issues raised by programming. Nevertheless, most of the major orientations now currently discussed were present in this book, especially the need to increase our knowledge of the cognitive processes underlying programming and its learning, and to define more accurate indicators of this activity and experimental investigations using tasks and subjects more representative of real programming.

Most experimental studies at this time were conducted by computer scientists, and very few referred to psychological theory or methodology. The purpose of these studies was rapid assessment of tools on the sole basis of finished products and no attempts were made to gain insights into the activity itself. This yielded a number of problematical and sometimes contradictory experimental results. A good example was the evaluation of flowchart use in comparison with listing, for which improvements (Wright, 1977) as well as lack of effect (Shneiderman *et al.*, 1977) had been shown on overall performances without understanding the reasons for this difference. However more detailed analyses of performance enabled researchers to precisely define the activity components affected by flowchart use (Brooke and Duncan, 1980; Gilmore and Smith, 1984): overall performance is shown to be improved only in situations where these activity components are crucial. Shneiderman (1980) provides a good review of the state of the art at this 'second generation' time.

When an experiment is not initially thought out within the framework of a theory, interpretation is seldom straightforward. As far as psychological processes are concerned, sole reference to a computer science theory can be misleading, especially when there are major disparities between computer science objects and programmer representations. Program complexity measures are particularly informative in this respect. A complexity measure is based on a certain way of structuring programs. In order to be psychologically valid for a certain kind of programmer, the structure must be consistent with the representational structures this programmer uses. For example, Halstead's metrics (1977) evaluate information transmitted by the operators and operands in the program. The structure which is considered here is the surface structure of the program, probably valid for beginners or professionals confronted with a very unknown program. As far as a deeper structure is concerned, this measure is inadequate since it ignores the high-level structures (chunks) used by expert programmers in understanding programs (Curtis *et al.*, 1984).

The 'third generation' of the psychology of programming was initiated in the early 1980s by a wide-ranging debate on the theoretical (e.g. see Hoc, 1983) and methodological (e.g. see Moher and Schneider, 1982) grounds for this kind of study. Pioneering efforts were followed by an increase in the number of psychologists studying programming, especially notational and debugging aspects (Green, 1980). At the same time, some computer scientists in the cognitive science field were developing cognitive approaches to programming that had a direct bearing on Psychology, e.g. Soloway *et al.* (1982), studied programming knowledge representations in terms of hierarchical schemas and plans.

The originality of these recent studies lies in a more indepth investigation of programming as an activity, through tools such as individual protocol analysis and cognitive modelling. These studies have benefited from enhancement by cognitive psychology methodology, which aims to elicit the externalization of covert behaviour. New experimental paradigms have been generated from psychological theories (on text comprehension, human problem solving and planning, etc.; see Chapter 1.4) and hypotheses drawn from observation (especially verbal report techniques; see Chapter 1.5).

The aim of the present volume is to explore this fast-growing trend in the psychology of programming. The value of psychological frameworks is stressed, but greater emphasis is placed on the need for a combination of psychology and computer science approaches. Computer science has developed languages, tools and environments that implicitly represent diverse conceptions of programming and enter into the definition of programming tasks (e.g. task requirements). Programming activity cannot be defined in isolation from this cultural environment. Computer science conceptions need to be assessed before investigating their role in programmer activity. A number of psychological difficulties can be anticipated by an *a priori* analysis of the coherence of these conceptions. Floyd (1984) provides this kind of analysis in her comparative evaluation of programming methods (e.g. decomposition of the programming process into sequential steps which turn out not to be independent and well defined when they are implemented).

The implication is that a psychological investigation must begin with analysis of these rapidly evolving concepts which range from very procedural approaches to functional, logical, object-directed, and more-declarative ones (see Chapter 1.1 by Pair). At the same time the number of styles of interacting with programs is rapidly increasing. Today's programmers have far more variety of possible approaches to choose between (see Chapter 1.2 by Green). Most of the studies referred to in this book deal with procedural programming, but very recent advances that are still unavailable to the general public are also presented. Pennington and Grabowski (Chapter 1.3) describe the richness of programming activity by defining its cognitive components, and their relationships, beyond program design, e.g. problem and program understanding, debugging, etc. An introduction to cognitive psychology is presented by Ormerod (Chapter 1.4) who selects the main concepts that are relevant to the psychology of programming and are used in the diverse chapters of this book. Gilmore (Chapter 1.5) introduces the reader to observational and experimental methodology in cognitive psychology and shows how scientific results are obtained.

The theoretical and methodological issues presented in Part 1 are complemented by Parts 2 and 3 which review the main research findings in two areas: features of programming languages and the learning of programming (Part 2), and expert skills and job aids (Part 3).

These studies have been run in more or less simplified situations, even though some are devoted to real programming tasks with a realistic level of complexity. The final part of this book addresses broader issues, related to everyday programming in companies.

Readers should measure the ecological validity of the data presented here in relation to the type of programmer the study deals with. Programmers, whether they are novices or experts, do not constitute a homogeneous population.

Novices belong to at least two distinct categories and their goals in the learning of programming differ. At school, programming is taught for enrichment, or in a transfer perspective which assumes that other knowledge can be acquired during programming learning (e.g. some understanding of mathematical concepts which can help the pupil in mathematics). A very limited amount of time is devoted to programming, which raises the issue of its true educational value (see Chapter 2.5). Adult and young adult novices can train to become professional programmers. Here training can be lengthy and cover several converging types of knowledge pertaining to coding in a programming language, using well-known algorithms, representing specifications in an efficient way for programming, etc. The relevance of teaching material to the working world is essential for assessing training curricula of this type.

Experts are not more homogeneous as a group, and can be defined by the regularity of activity, and the level and nature of expertise.

Activity can be regular or casual. Regular programmers are those who spend their time designing, coding, debugging, documenting and modifying programs. Casual programmers are people whose main activity is not programming. They work in other fields and they program for professional purposes. In terms of training as well as job aids, the needs of these two kinds of programmers are quite different. Few studies have been devoted to casual programmers who are certainly in the majority and who require intelligent support systems, such as advice giving systems (e.g. see Giboin, 1988).

Expertise can be general or specific. General programmers need broad expertise in languages, tools, methods, team work, etc. They may be called upon to deal with almost any kind of programming environment or problem domain during their careers. They have to learn to use multipurpose tools and produce high-quality software. More-specialized programmers are expert in a particular application domain, such as management, statistics, programmable controller programming (see, for example, the work done by Visser (1987) referred to in Chapter 3.3), etc. For this type of programmer, specific-purpose languages and tools can be developed and taught.

References

Brooke, J.B., Duncan, K.D. (1980). Experimental studies of flowchart use at different stages of program debugging. *Ergonomics,* **23**, 1057-1091.

Curtis, B., Forman, I., Brooks, R., Soloway, E., Ehrlich, K. (1984). Psychological perspectives for software science. *Information Processing and Management,* **20**, 81-96.

Dahl, O.J., Dijkstra, E.W., Hoare, C.A.R. (1972). *Structured Programming.* London: Academic Press.

Floyd, C. (1984). *A Comparative Evaluation of System Development Methods.* Berlin: Technische Universität Technical Report.

Giboin, A. (1988). The process of intention communication in advisory interaction. *3rd IFAC Conference on Man-Machine Systems, Oulou (Finland).*

Gilmore, D.J., Smith, H.T. (1984). An investigation of the utility of flowcharts during computer program debugging. *International Journal of Man-Machine Studies,* **20**, 357-372.

Green, T.R.G. (1980). Programming as a cognitive activity. *In* H.T. Smith and T.R.G. Green (Eds), *Human Interaction with Computers.* London: Academic Press.

Halstead, M.E. (1977). *Elements of Software Science.* New York: Elsevier.

Hoc, J.M. (1983). Psychological study of programming activity: a review. *Technology and Science of Informatics,* **1**, 309-317.

Hoc, J.M. (1988). *Cognitive Psychology of Planning.* London: Academic Press.

Miller, L.A. (1981). Natural language programming: styles, strategies, and contrasts. *Perspectives in Computing,* **1**, 22-33

Moher, T. and Schneider, G.M. (1982). Methodology and experimental research in software engineering. *International Journal of Man-Machine Studies,* **16**, 65-87.

Newell, A. and Simon, H.A. (1972). *Human Problem Solving.* Englewood Cliffs, NJ: Prentice Hall.

Rouanet, J. and Gateau, Y. (1967). *Le Travail du Programmeur de Gestion: Essai de Description.* Paris: AFPA-CERP.

Shneiderman, B. (1980). *Software Psychology: Human Factors in Computer and Information Systems.* Cambridge, MA: Winthrop.

Shneiderman, B., Mayer, B.R., McKay, D. Heller, P. (1977). Experimental investigations of the utility of detailed flowcharts in programming. *Communications of the ACM,* **20,** 373-381.

Soloway, E., Ehrlich, K., Bonar, J. and Greenspan, J. (1982). What do novices know about programming? *In* A. Badre and B. Shneiderman (Eds), *Directions in Human-Computer Interaction.* Norwood, NJ: Ablex.

Visser, W. (1987). Strategies in programming programmable controllers: a field study on a professional programmer. *In* G. Olson, S. Sheppard, and E. Soloway (Eds), *Empirical Studies of Programmers: Second Workshop.* Norwood, NJ: Ablex.

Weinberg, G.M. (1971). *The Psychology of Computer Programming.* New York: van Nostrand.

Wright, P. (1977). Presenting technical information: a survey of research findings. *Instructional Science,* **6,** 93-134.

Chapter 1.1

Programming, Programming Languages and Programming Methods

C. Pair

Centre de Recherche en Informatique de Nancy, BP 239, 54506 Vandœuvre, France

Abstract

A 'program' has meant many diverse things at different times. The oldest view is that programming is describing calculations; starting with the early languages such as Fortran and Basic, improved languages have been developed, in which the difficult GOTO constructions have been avoided and a method of top-down structured programming has been supported. The top-down method has certainly increased the safety of programs, but it has never given a clear description of how calculations should be broken down into smaller problems. A second view presents programming as defining functions. A program is a chain of functions which each build an object intermediary between input and output, and the act of programming becomes a matter of progressively enlarging a library of functions. A number of versions of functional programming are sketched. In the most recent view, programming is presented as defining and treating objects, which combines the two previous views: breaking down the calculations to obtain an algorithm, and representing intermediary objects. A program defines inter-relationships between objects, and one new style of programming is progressively enriching not just an unstructured library of functions, but a universe of objects with well-defined relationships.

Psychology of Programming
ISBN 0-12-350772-3

1 Introduction

At first sight programming seems a very straightforward activity; however it is always deceptively so. Debugging programs takes time, there are always errors and it is a painstaking task to track them down and correct them. Programmers have all experienced this; if they have not found any solutions as such to the problem, they have nevertheless worked out ways of playing it safe. The psychology of programming (Hoc, 1982) studies these strategies.

The computing community also has its own collective history on this subject. At the end of the 1960s, a crisis emerged in software, programming and languages. Methodology, top-down design, structured programming, functional programming were mentioned for the first time. Twenty years later these topics continue to be discussed. New types of languages – new compared to Fortran or Basic – have appeared or have been rediscovered: Lisp, Prolog, Smalltalk, etc. Will they solve all the problems? That remains to be seen. If one is to come to terms with the issues at hand, it is necessary to define what a program is. Indeed, the idea of what a program is inevitably affects programming, the more so as it is reflected in the languages as well as in the methods that are available and are in use in education, and in the world of programming aids.

This chapter aims to point out the prevailing conceptions of the notion of 'program' in computer science, the way they are worked out, and the languages and the methods they have generated.

2 What is a program?

The first answer is *syntactical*: a program is a text constructed according to certain grammar rules. This point of view has long prevailed both in teaching (teaching the forms of instructions one after the other, instructing where not to forget a comma, and so on) and in research (the first well-written chapter of computer science was probably the theory of languages, on which Chomsky had a more lasting influence than in the study of natural languages; the compilers are organized around the syntactical analysis, etc.). In fact, the syntactical viewpoint remains predominant for beginners – it is a necessary stage, and cheaper training ends there.

It is obviously not enough. It is useless to know a language if one does not understand what it expresses, the meaning of its sentences, its *semantics*. So the question is: what does a program express?

An opinion poll conducted on this subject would probably reveal that in the eyes of most people a program describes a calculation. Programming, we know, came into existence at the same time as the computer, and the one aim of that tool is to calculate. Of course the word 'calculate' has to be taken in a wider sense: printing, drawing graphs, interpreting data transmitted by sensors, giving orders to a robot, consulting a dictionary or a file, all these are forms of calculation. It is therefore possible to define a calculation as a sequence of changes in the state of the machine (and particularly of its memory).

In fact, it is not correct to say that a program describes *one* calculation; in the majority of cases the calculation worked out is based on external inputs. So a program generally expresses a whole *set of calculations*, most often infinite – or else a function linking a calculation to each possible input.

But of what use are these calculations, and what do they express in themselves? It is sometimes the case that the calculation is the end in itself, for example, if it controls cartoons, or a game, or (more rarely) a robot; i.e. one is interested in all or some of the stages through which the machine, or a mechanism it controls, passes. But mostly it is not the calculation that is important but its result. The person using the computer is not really interested in the stages; what he is interested in is the outcome.

From the customer's point of view, the program leads from inputs to results; it expresses – to use a mathematical expression – a function : input → result.

This then is a second model of the notion of program. If the first one – the set of calculations – can be called *imperative* or *procedural* (because it expresses *how*) this second model can be called *declarative* or *definitional* (because it expresses *what* – what one expects, what has been asked for or specified).

The choice between these two points of view definitely influences the activity of programming. Programming consists in transforming a *specification*, which describes a function, into a program, that is to say, a text which can be interpreted by a machine in order to calculate this function. Does proceeding from the specification of the problem to the program involve a mental image of all the calculations, a mental execution strategy (see Hoc, 1983; Chapters 3.1 and 3.3)? Or is it possible that only definitions of functions are involved?

It would seem, indeed, when one watches most beginners (at any rate), that one can answer the first question positively and, therefore, the second one negatively: possibly because the model is dictated by the tool; or possibly, on the contrary, because the programmer refers to the way a human being would work it out; or because the notion of function seems abstract. This is not without drawbacks and leads to errors, notably because it is not easy to have a mental image not only of *one* calculation but of a generally infinite *set* of calculations.

It is possible to react in two ways to this discovery, and this has been done by researchers and programming tutors:

i. to accept going through *all* the calculations and to see how best to help it run smoothly;

ii. or to refuse it and try to train people to avoid it.

In both cases tools are used: languages, methods, software aids.

3 Programming is describing calculations

Historically, this is the first way as it is the closest to the tool: a calculation is a sequence of changes of state. Each possible change of state will therefore be described by an instruction (modify the value of a memory word), and a calculation by a series of instructions. This is how 'machine languages' operate. In fact, it is not one calculation that has to be described, but a set of calculations which is generally infinite and which yet has to be described in a finite manner. To this end one invents the jump (or *go to*); it must be *conditional* to be able to give a good description of an (*infinite*) set of finite calculations and not a single infinite calculation.

The first algorithmic languages (Fortran, but also later Basic) are not founded on very different ideas, the only difference being that the memory words are represented by symbols: variables if they contain values, labels if they contain instructions.

Programming generated by these conceptions consists in considering the first instruction of the calculation, then the second, then the third. . . and in saying from time to time: at this point let's start again (see Chapter 2.4). It is programming at 'grass roots', setting everything at the same level; however, this is quite normal for beginners since it transposes the way of performing the task 'by hand'. A flowchart expresses it well.

Computer scientists came to realise, however, that jumps produced errors. It is easy enough to understand why: one is thinking of a calculation and, suddenly, there is a change in levels to designate a point in the program; what is more, the link with the described calculations becomes obscure as soon as the references cross (see Green, 1980). There is also collision in this case between the semantic aspects (calculations) and the syntactic ones (point in the program), and it is well known that such collisions are the source of misunderstandings and paradoxes; for instance, the one concerning 'the first word of the English language, in alphabetical order, which cannot be defined in under twenty words', but which has just been defined in eighteen.

In short, towards the end of the 1960s the programming reformers (Dijskstra, Hoare, Wirth *et al.*) rejected in a sometimes slightly dogmatic manner the *go to*; and the more modern algorithmic languages (Pascal, Ada, etc.) have turned it into an object of secondary importance by providing ways of dispensing with it.

Which ways? Ways which make it possible to describe a calculation, or rather a set of calculations, differently than at the 'grass roots' level, instruction after instruction. So the calculations have to be broken down into parts which are given names:

program: part 1; part 2; part 3

The next stage consists in breaking down the parts, on and on, until one reaches elements which are immediate to program: in this way a structured tree-shaped analysis is obtained. With this method (*top-down* or *structured programming*, see Chapter 3.3) details are only meant to be considered progressively. And within the program, each part can give rise to a procedure, particularly if the language allows for one procedure to be inserted into another.

This step-by-step breaking down procedure, however, only allows for the description of a set of calculations if a conditional statement is introduced:

if *condition* **then** *part 11* **else** *part 12*

To solve the problem of describing infinite calculations, two fairly equivalent tools are available:

* iteration, of the type:

  while *condition* **repeat** *action*

  (more powerful than Fortran's **do** where the number of repetitions is fixed);

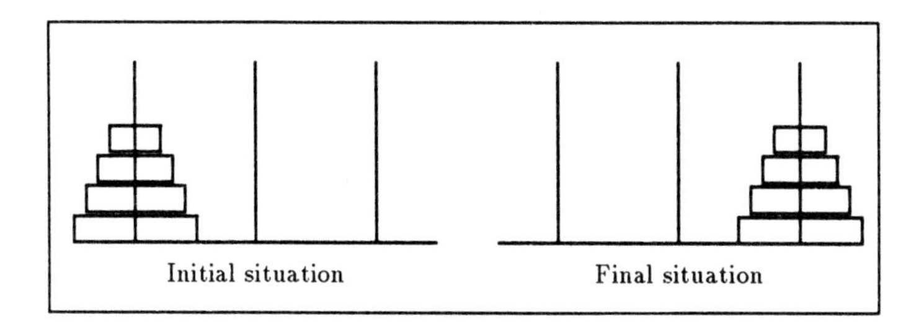

Figure 1:

* recursivity, which appears as a natural consequence of the top-down decomposition process: one stops decomposing not only when one reaches an 'elementary' part or an already known and programmed part, but also when one comes upon the very problem that one is programming.

When, however, this recursivity is presented in the context of the imperative top-down programming and of the algorithmic languages, it seldom goes down well and one is left with a feeling of uneasiness (see Chapter 2.4). One of the reasons is that one generally does not find the exact problem which is being analysed, but a similar one, with slightly different data. So the two problems need to be unified into a more general problem by the introduction of parameters. This makes the description more complex and changes its level.

Besides, when writing a recursive procedure, is one in fact still giving a description of calculations? Let us examine, for example, the well-known problem of the Hanoï towers: the goal is to move n pieces from a *starting* post to a *finishing* post using a third post as an *intermediary* (see Figure 1); at each step the top piece of one of the posts is moved, but may not be set onto a smaller one. The recursive analysis consists in writing that the goal can be reached by:

* passing $n-1$ pieces from the *starting* post to the *intermediary* post using the *finishing* post (similar problem);

* then moving the number n piece from the *starting* post to the *finishing* post;

* finally passing the remaining $n-1$ pieces from the *intermediary* post to the *finishing* post using the *starting* post.

The procedure obtained is the following:

```
procedure h (n, starting, finishing, intermediary);
          if n > 0 then h (n-1, starting, intermediary, finishing);
                        move (n, starting, finishing);
                        h (n-1, intermediary, finishing, starting)
```

But does this not amount to a definition of the set of calculations, by a property defined as a function of a certain number of arguments?

Even more so than for procedures, the question is crucial for the programming of functions; for instance, to calculate rapidly a power a^n, it is possible to use the fact that it is the square of $a^{n/2}$, at least if n is an even number; more precisely, one obtains the function:

```
function p(a, n) : if n = 0 then p := 1
                   else if even (n) then p := p (a, n, div 2) ≠ 2
                   else p := p (a, n, div 2) ≠ 2 * a
```

(*div* yields the integral quotient and $\neq$ stands for the raising to a power).

This function definition is an *equation* verified by the function p:

```
p(a, n) = if n = 0 then 1
          else if even (n) then p (a, n div 2) ≠ 2
          else p (a,n div 2) ≠ 2 * a
```

In fact the notion of function in algorithmic languages is a hybrid: describing calculations or yielding a value? The fruit of this strange coupling is the famous 'side effect' which, during the computation of a function, also modifies the state of the memory.

Finally, it must be added that even if structured programming has represented an important step towards greater safety, it does not provide any lead concerning the central question: how does one break down a problem?

4 Programming is defining functions

In many cases, when one examines the breaking down of a problem in structured programming, one notices that each part builds an object, intermediary between input and result.

Consider the example of linear regression: input, values of n experimental measures for variables $x_1, ..., x_p$ (explaining variables) and y (explained variable); output, the coefficients $b_0, b_1, ..., b_p$ expressing by the least-squares method y as a linear combination $b_0 + b_1 x_1 + ... + b_p x_p$.

The mathematical study shows that the calculation can be done in the following manner:

construct the matrix M of the measurements $(x_{ij} y_i)$ with $x_{i0} = 1$

calculate the transposed T of the partial matrix obtained by removing the last column from M

calculate the product P of T by M

triangulate the linear system of matrix P

solve the system

write the coefficients

Or more precisely:

$M =$ *construction (input)*

$T =$ *transpart (M)*

$P =$ *product (T, M)*

$A =$ *triangulation (P)*

$B =$ *solution (A)*

result $=$ *writing (B)*

In fact these are *definitions* obtained by introducing *functions* which have to be described as well. It is actually possible to give these definitions in an arbitrary sequence (for example, by starting from the result and introducing the intermediaries which seem useful), since a calculation sequence automatically emerges from the linking of intermediaries; consequently it does not need to be specified. But the notion of calculation is altogether forgotten: the style obtained is purely algebraic. This can be referred to as *declarative* or *functional programming*.

The earliest attempt at promoting this style of programming through a language is Lisp (McCarthy *et al.*, 1962; McCarthy, 1978), a language from the same period as Fortran, and long before Pascal. Although knowledge of it for a long time was not widespread, perhaps because of its syntax and its rank obstinacy to do everything starting from a very small number of concepts, and also because of the few applications requiring at that time such programming; but it has now become, thanks in particular to its applications in artificial intelligence, one of the prominent programming languages. It must be mentioned, however, that pure Lisp did not survive for long and that variables, iterations, etc. were soon introduced again, but as elements of secondary importance.

Functional programming urges one, even more so than top-down imperative programming, to come back to functions that have already been programmed. What matters is less writing a large program than progressively enlarging a library of functions introduced in the most logical sequence possible: the distinction between programming and controlling a system then disappears.

Other attempts to combine the functional style of analysis with algorithmic programming (particularly concerning the notion of iteration) can be mentioned: e.g. Lucid (Ashcroft and Wadge, 1976) and deductive programming (Ducrin, 1984).

J. Backus's ideas are more radical: the idea is no longer as with Lisp to define a function by giving the expression of its result (see function p at the end of Section 3) but to consider it rather as constructed by various types of composition from bricks, i.e. elementary or predefined functions: function p, for example, would be defined as a conditional composition involving the predicate of equality, the constant functions 0 and 1, the *even* predicate, the *square, div, multiplication* functions. It should not come as a surprise (all one is doing is, in fact, composing a program starting from library programs), but it is nevertheless a shock to our present mathematical culture, for, if Lisp does away with the notion of the computing variable of Fortran and is satisfied with mathematical variables, Backus goes as far as abandoning this concept. It is in fact the point of view of combinatory logic (Curry and Feys, 1968).

Yet another way of seeing the definition of a function is to describe a relation between its data and its result. The greatest common divider of two integers, for example, is defined by

$$y = gcd\ (a,\ b) \Leftrightarrow divides\ (y,\ a)\ and\ divides\ (y,\ b)$$
$$and\ \forall z\ (divides\ (z,\ a)\ and\ divides\ (z,\ b) \Rightarrow divides\ (z,\ x))$$

The language used here is that of mathematical logic, more precisely that of first-order predicate calculus. And it makes it possible to describe more general relations than functions:

$$uncle\ (x,\ y) \Leftrightarrow \exists z\ (brother\ (x,\ z)\ and\ parent\ (z,\ y))$$

This type of language can be used for the specification of problems, as was done by Abrial in the Z language: in this sense, specification is the first step towards a program. But it is also possible to go no further than specification and to shift away from the description of a calculation by limiting oneself to defining, in this type of logical language, the relations between arguments and results, leaving it up to interpretation software to discover the calculations that will lead to the result. These calculations are in fact reasonings. This is what happens in expert systems. It is also the idea put into practice by the Prolog language (Roussel, 1975; Colmerauer *et al.*, 1983). In theory, the author of the program would not need to know anything about the manner in which the interpreter will draw the deduction which leads to the results. In theory at least – reality is not so straightforward.

5 Programming is defining and treating objects

The above is flawed by a serious defect. Stressing the breaking down of calculations or the definition of functions and relations leads one to forget that calculations as well as functions deal with objects: objects mentioned in the specification of a problem; objects processed by the machine or, at a slightly higher level, directly accessible in the programming language. The latter can be numbers, strings, arrays, files; the former customers, parts, graphs, polyhedrons, logical formulæ. . . Programming is also the transformation from the latter to the former; in other words, their representation.

There are, therefore, two aspects to programming: breaking down the calculation to obtain an algorithm and representing objects. One way of dealing with this duality – the one generally adopted by beginners – consists in concentrating first and foremost on the representation of the objects, because this leads them back to a more familiar situation: treating a problem bearing on the objects of language. In fact this method is induced by most languages, and Lisp more than any other, since the types of objects handled are particularly poor: there is only one, *lists*, or more precisely, *trees*; it is true that starting from there it is possible to describe all other types, but it is not necessarily easy.

Taking successively into account the two dimensions of programming – the representation of the types then the breaking down of the calculation – does not provide any guide as to representation and one can hardly expect to achieve an efficient representation for the problem in hand. So it is better to start off with the breaking-down process, and then to choose representations which will make it possible to work out efficiently the functions and procedures brought to light by the breaking down, a process which can be repeated at several levels. For example, it is very important that the characteristics of an object which are useful in the algorithm be directly accessible in its representation: they can be fields of a record representing the object (Pair *et al.*, 1988).

This leads one to think that what matters most of all is not the objects as such, but the operations executed on them. This idea joins up with similar viewpoints in other areas: mathematics, epistemology, psychology, and even technology. In computer science, it has given rise to the notion of abstract data type: contrary to the viewpoint generated by the classical algorithmic languages, here a type is essentially characterized by its operations and to define a type is to give a packet of operations involving it; they will guide its representation. The first language that adopted this viewpoint must have been Clu (Liskov, 1975), and these ideas have since been incorporated into Ada (Ledgard, 1980).

Taking this approach, however, one soon realises that types of objects are related, particularly in a hierarchical manner: a rectangle is a kind of quadrilateral and a square is a kind of rectangle, which means that to define the class of rectangles, and the operations which will be carried out on them, it will be possible to refine the notion of quadrilateral without repeating everything that can characterize a quadrilateral. It should also be noticed that this hierarchy between types is not in itself different from that which connects one object with its type: a type shows properties valid for all its objects, each one of them being a particular variation of it. It is also possible to link it to psychological notions such as the prototype, i.e. an object of a type from which others can be obtained by modifying certain characteristics. This can be seen as a tendency to approximate a 'naive' logic, whereas functional programming leans more towards mathematics.

In programming, the hierarchy between types presents similarities with the hierarchy between problems and subproblems or between procedures. In the latter case, there exists a rule concerning the area of validity of the identifiers in the interleaved procedures, which allows for understatements. And the duality of viewpoints – processing and objects – allows two hierarchies, which are not generally identical, to coexist. An algorithmic language such as Pascal clearly favours the hierarchy of processing: in particular a type declaration is separated from the functions and procedures which should accompany it.

It is possible to adopt the opposite point of view and favour the notion of object. There, programming is seen as defining objects. It means enriching a universe of objects; and an object is described from other objects (e.g. its type) by specifying certain characteristics (a value, for instance).

This viewpoint generates yet another new style of programming: progressively enriching not only an unstructured library of functions, but a universe of objects in relation with one another.

Even if these ideas appear relatively new, they already have a long history – part of which can be found in Simula (Dahl and Nygaard, 1966), an adaptation for the simulation of the algorithmic language Algol 60 (Naur, 1960). They are the basis of the object-oriented languages of which Smalltalk (Kay and Goldberg, 1976) is the best known: 'object-oriented languages' not because others would disregard the objects, but because the programming process in this case is guided by the objects, their definitions and their relations.

6 Conclusion

What about the future then? Convergences appear. Object-oriented languages are generally created starting from Lisp, which becomes a 'machine language' of functional programming, on which more sophisticated languages from the viewpoint of represented objects are built. Prolog also handles a single type of object – trees again – so it would be useful to improve it by diversifying the types; that is why a certain number of attempts are being made to bring logical programming into convergence with object-oriented languages. This is because artificial intelligence continues to play a driving role in the evolution of programming and of languages. However, there will be no true artificial intelligence if data bases are not included; and one discovers here a whole universe of objects and relations from which to carry out reasoning.

On the other hand, even if programming through a mental execution strategy appears theoretically as a useless detour and a source of errors, experience shows that it is difficult to do away with it completely, and for many people the notion of calculation seems more concrete than the mathematical languages to which the declarative viewpoint leads. Object-oriented programming may render this discussion useless by working on objects close enough to the problem set for the analysis to be very simple and then doing a series of representations.

One must make distinctions, not only according to the kinds of application, but – and this is new – according to the programmers. The language used is another dimension. Formerly one used to say that the language in which a program was written was not very important: all languages used were indeed once similar, all were algorithmic languages (except Lisp which was not well known). This is no longer the case nowadays.

The development of programming languages and methods, and the teaching of them, have up to now hardly been linked to a psychological study of the activity of programming, and this can account for certain failures. To be of any use, however, psychology must go beyond the procedural aspect of programming; it must take into account those other styles which, even if they are not new, are becoming more and more important nowadays due to the variety of applications and the training that programmers receive.

References

Ashcroft, E.A. and Wadge, W.W. (1976). Lucid, a formal system for writing and proving programs. *SIAM Journal of Computing* **5**, 336-354.

Backus J. (1978). Can programming be liberated from the von Neumann style ? A functional style and its algebra of programs. *Communications of the ACM,* **21**, 613-641.

Colmerauer, A., Kanoui, H. and Van Caneghem, M. (1983). Prolog, bases théoriques et développements actuels. *Technique et Science Informatiques* **2**, 271-311.

Curry, H. B. and Feys, R. (1968). *Combinatory Logic,* vol. 1. Amsterdam: North Holland.

Dahl, O.J. and Nygaard, K. (1966). Simula, an Algol based simulation language. *Communications of the ACM,* **9**, 671-678.

Dijkstra, E.W. (1976). *A Discipline of Programming.* Englewood Cliffs: Prentice Hall.

Ducrin, A. (1984). *Programmation.* Paris: Dunod.

Green, T.G.R. (1980). Ifs and thens: is nesting just for birds? *Software Practice and Experience,* **10**, 371-381.

Hoc, J.M. (1982). L'étude psychologique de l'activité de programmation: une revue de la question. *Technique et Science Informatiques,* **1**, 383-392.

Hoc, J.M. (1983). Analysis of beginner's problem-solving strategies in programming. *In* T.R.G. Green, S.J. Payne and G. van der Veer (Eds), *The Psychology of Computer Use.* London: Academic Press, pp. 143-158.

Kay, A. and Goldberg, A. (1976). *Smalltalk 72, Instruction Manual.* Palo Alto: Xerox Research Center.

Ledgard, H. (1980). *Ada, an Introduction, Ada Reference Manual.* New York, Heidelberg, Berlin: Springer-Verlag.

Liskov, B.H. (1975). An Introduction to Clu. *In* S.A. Schuman (Ed.), *New Directions in Algorithmic Languages.* Rocquencourt: INRIA, pp. 139-156.

Liskov, B.H. and Zilles, S.N. (1974). Programming with abstract data types. *SIGPLAN Notices,* **9**, 50-59.

McCarthy, J. (1978). History of Lisp. *SIGPLAN Notices,* **13**, 217-223.

McCarthy, J., Abrahams, P.W., Edwards, D.J., Hart, T.P. and Levin, M.I. (1962). *Lisp 1.5. Programmer's Manual.* Cambridge, MA.: MIT Press.

Naur, P. (1960). Report on the Algorithmic Language Algol 60. *Nümerische Mathematik,* **2**, 106-137.

Pair, C., Mohr, R. and Schott, R. (1988). *Construire les algorithmes.* Paris: Dunod.

Roussel, P. (1975). *Prolog, manuel de référence et d'utilisation.* Groupe d'intelligence artificielle. Marseille: Université de Marseille.

Wirth, N. (1976). *Algorithms + Data Structures = Programs.* Englewood Cliffs: Prentice Hall.

Chapter 1.2

The Nature of Programming

T. R. G. Green

MRC Applied Psychology Unit, 15 Chaucer Road, Cambridge CB2 2EF, UK

Abstract

'Programming' is an exceedingly diverse activity, and many questions confront anyone who tries to say how programmers should work. This chapter attempts to describe some of the issues where cognitive psychology is relevant, without offering answers. Different programming cultures stress different virtues, on one hand neatness and well-definedness, on the other hand openness and effectiveness. The neat-scruffy differences show themselves both in individual styles and in claims that languages should be small, or that languages should contain all the necessary tools; although the argument ultimately depends on judgements of utilities, there are many cognitive issues not yet answered. The cultural differences are supported by various social mechanisms whose study would be of interest in its own right. Programming environments show similar divisions into neat or scruffy, but they also differ in being high or low-level. Low-level text-based environments are still commonest but structure-based editors, usually built on syntactic structure, are becoming more common. More recently, editors based on semantics have been investigated. Each higher level poses more cognitive questions, most of which remain unanswered. Developments in visual representations of programs also raise urgent questions of what information should be displayed and how, as do the increasing numbers of specialist languages. The chapter ends with a sketch of some of the most promising recent developments, indicating yet again a few of the cognitive issues that are raised.

Psychology of Programming
ISBN 0-12-350772-3

1 Introduction

This is a chapter about the diversity of tools used by different programmers, and the differing traditions and enterprises. Obviously, only thumbnail sketches can be given. The point to be made is the incredible diversity of ways to program, and the large number of unanswered questions that arise. The contributions to the psychology of programming that are reviewed elsewhere in this book can only be properly evaluated by appreciating the field as a whole.

2 The cultures of programming

Although many different programming languages have been devised for many different purposes, it is intriguing to see that the same schools of thought constantly are reborn in new areas. And it is discussions between these schools of thought that make passions fly highest!

2.1 'Neat' or 'scruffy'?

One argument crops up again and again, epitomized in the design of the two well-known languages C and Pascal. C allows the user to get at the internals of the machine and to control individual registers. It also allows the user to make the most horrible mistakes. For instance, the bounds of arrays and other data structures are not checked at runtime, so that users can access non-existent elements and can process them as though they were real data. Error messages are short and often hard to interpret; C frequently gives up when a runtime error occurs, makes a core dump, and quits, leaving the user to work out the mistake from first principles or else to poke through the core dump with a disassembler. C also has a very terse syntax which is not completely unambiguous, and which was not originally defined by a formal grammar; users can modify the syntax to some degree by using 'def' statements, which can certainly clarify programs at times but which also allows people to play games and write deliberately hard programs (Figure 1).

In contrast Pascal keeps the user away from the internals of the machine, and protects against as many mistakes as possible. Mistakes are detected at compile-time whenever possible, and further checks are made at runtime. The syntax was designed to be unambiguous and was defined precisely by a context-free grammar.

Neat languages are *well defined*: a single document fully defines the language and another document defines the source program. The behaviour of the program does not depend on recent history, such as what modifications have been loaded into the system, and as far as possible the meaning of any part of the program can be ascertained without reference to any other part ('no side effects'). Scruffy languages may not have been given a full definition, and sometimes one can only discover exactly how a particular statement is executed by observing how the compiler behaves (C was originally like that), while their interpretation very frequently depends on how the system has been tailored for local purposes.

Neat languages actually prevent programmers from doing things that might be 'dangerous' – which is why Pascal makes it hard to get at individual registers in the machine. Scruffies regard that as a paternalistic, even authoritarian, attitude, and programmers in scruffy languages are expected to look after themselves.

```
  #define o define
  #o ___o write
  #o ooo unsigned
  #o o_o_ 1
  #o _o_ char
  #o _oo goto
  #o _oo_ read
  #o o_o for
  #o o_ main
  #o o__ if
  #o oo_ 0
  #o _o(_,__,___) (void)___o(_,__,ooo(___))
  #o __o
(o_o_<<((o_o_<<(o_o_<<o_o_))+(o_o_<<o_o_)))+(o_o_<<(o_o_<<(o_o_<<o_o_)))
  o_() {_o_ _=oo_,__,___,____[__o] ;
  _oo ______;_____:___=__o-o_o_;
  _______:_o (o_o_,____,__=(_-o_o_<___?_ o_o_:___)) ;
o_o(;__;_o(o_o_,"\b",o_o_),__) ;
  _o(o_o_," ",o_o_); o__(--___)_oo ______; _o(o_o_,"\n",o_o_);
  _______:o__(_=_oo_(oo_,____,__o))_oo _____;}
```

Figure 1: Delightful ingenuity has been used to create a C program that is perfectly correct and deliberately unreadable. It comes from a competition where many other extraordinary programs were submitted. The trick used here is to define the string 'o' as being a replacement for the string 'define'; thus the second line 'really' says `#define ___o write`, etc.

Pascal and C, so different in every respect, are near extremes of these two philosophies. But there are many other similar examples: among the scruffy languages are APL, Forth, Lisp and Fortran; among the neat languages are Algol 68, Occam, Scheme and Eiffel. The differences are not just in language – other arenas, notably programming environments, are mentioned below – and nor are they absolute: good programmers can choose to program neatly in a scruffy language, for instance. Moreover the needs of the task may dictate a choice based on trade-offs; one wants very secure programs for situations where mistakes are costly (and the costs of mistakes today are much greater, from beaching a supertanker to creating a nuclear 'incident'), and one needs equally to get at the details of the machine level for time-critical work such as real-time high-bandwidth communications. Yet personal style continues to play a large part, and claim and counter-claim can sometimes be acrimonious.

Cognitive psychology cannot pretend to dictate answers in such a deep dispute. But both sides have based their claims partly on a theory of cognitive processes, and what we psychologists *can* do, and occasionally have done, is offer better theories, based on real data rather than intuition. Some of these are documented later, in Chapter 2.2 on 'Programming Languages as Information Structures'.

2.2 Individual style

The contrast between neats and scruffies certainly depends in part on the needs of the task and on the local culture, but it also depends in part on individual preference. Some people enjoy writing cryptic code, while others enjoy writing self-evident code. The mystique of cryptic programming, in which programs become fiendish puzzles for outrageously intelligent technocrats, was seriously dented in the 1970s by a series of polemics propounding the ideals of good style, such as Kernighan and Plauger (1974). This book contained a great deal of common sense; some of it was apparently wrong when tested in the laboratory – but all the same, it added up to a convincing case for straightforward, simple programming. Ingenuity for its own sake became less popular.

Despite that shift away from the baroque, individual style is still detectable in programming. What do we know about it? Next to nothing. Aptitude tests, widely investigated as a personnel selection tool, tell us nothing about programming style, and personality factors seem to have been unhelpful in investigations. One of the few successful investigations into personality factors in this area, perhaps even the only one, is the demonstration that risk aversion plays a part in success with certain types of programming environment, notably structure-based editors (Neal, 1987).

2.3 The urge to economy

Another type of contrast separates big, rambling programming languages from small, economical ones. Small languages using uniform principles are 'obviously' easier to learn and use, some say; others deny that, asserting instead that if a programmer needs facilities for, say, string handling, they should be provided, rather than having to be built from scratch each time. This is one of those disputes where each side has got hold of a truth, but equally each side refuses to see that the other side also has a truth. In consequence, I have heard some spirited arguments– and some incredibly silly claims. In general, neats tend to go for small languages, scruffies for big ones.

The basis of the 'small is better' argument is perfectly clear and intuitively appealing. Unfortunately it leads to absurdities. For example, in one version of Prolog the standard way to perform subtraction is by inverse addition. Addition is performed by writing

```
SUM (X, Y, Z)
```

with the effect that if `X` and `Y` were 'bound' (i.e. had already been given values) and `Z` was unbound, `Z` would be set to `X + Y`. No problem so far. But subtraction – setting `Z` to `X − Y`, say – is performed by writing

```
SUM (Y, Z, X)
```

This has two difficulties. It is hard for the programmer to work out how the terms should be arranged to generate the right sum, and it is hard for the reader to discover which is meant, addition or subtraction; that can only be discovered by working out whether the variables are bound or unbound at the time of execution. Here we see an interrelationship with the question of mental models of program behaviour: a possible reply, in the case of Prolog, would be that the programmer only 'ought' to need to know that the relation `Y + Z = X` holds, without needing to know how that relation is evaluated. Empirical research casts considerable doubt on that claim, however. In any case similar absurdities arise in other contexts; Green (1980, p. 285) points out that 'small is better' predicts that it would be better to use the single logical operator `P/Q`, meaning 'at least one of `P` and `Q` is false', rather than the usual untidy collection of **`AND`**, `OR`, etc. But if you do that, then straightforward expressions turn into monsters:

`(p or q) and r` *becomes* `(((p/p)/(q/q))/r)/(((p/p)/(q/q))/r)`

This is neither comprehensible nor beautiful.

2.4 'Natural' programming

The inventors of new techniques of programming frequently assert that their technique is better because it is 'natural'. They rarely say what they mean by that, let alone discuss why their invention is more natural than other available techniques! Are these claims well founded? Is there a programming language that is natural? In my opinion, no. One day, maybe, but not yet.

For example, one school of thought supports logic-based programming, on the grounds that our natural mental model is supposed to concentrate on logical relationships rather than on the order of executing actions. The best-known logic-based language is Prolog, which has been extensively studied as a possible teaching language for children and for students. But, in reality, many difficulties arise in Prolog, some of which reflect the simple fact that logic may be logical, but it is not natural; it is not how people think. Empirical research indicates that Prolog novices find execution-based models of computation easier than logic-based models (see chapter 2.5), and that even experts use both types of model, not just logic.

Virtually identical claims have been put forward for object-oriented programming: it is supposed to simulate in some important fashion how we naturally conceive the world – in this case, however, we are supposed to conceive the world not as logical relationships, but as objects and classes of objects, communicating by doing things to each other ('sending messages'). Yet the interrelationships between real-world objects are far more subtle and various than the interrelationships that are available in any programming system, and typically a set of objects (in the programming sense) has to be defined with very great care and skill to achieve the right effect. Both anecdotal evidence and observational data (Détienne, 1990) show that programmers have difficulty in deciding which logical entities shall be represented as objects and which as attributes of objects. Object-oriented programming may be effective, but it is certainly not artless: and in that case, it is not natural.

2.5 The social maintenance of culture

The differences between programming cultures are neither accidental nor short lived. Yet in many cases they seem to be independent of the language itself; for instance, Pascal textbooks usually use rather long identifiers, while C textbooks use rather short ones. How are these cultural differences maintained? There are several mechanisms.

Firstly, the pedagogic traditions are very different. A cursory examination of textbooks dealing with neat languages will reveal the prevalence of certain tutorial examples, such as the 'eight queens' problem and sorting and searching problems, which are very clean and well defined. Each problem is usually solved as a complete enterprise, rather than being the basis of evolutionary growth, and there is frequently some analytical reasoning about the invariants of the program or about its performance with respect to size of data set. Students are exhorted not to start coding until they have fully analysed the problem, nor to approach the computer until the coding details are fully worked out. Ideally, the code should run correctly first time. Code that does not solve the problem is an 'error'. The 'scruffies' work with a larger problem repertoire, and frequently prefer to give space to domain-specific problems, such as interacting with the operating system, or algorithms for natural-language parsing, rather than to analytical reasoning. Students are expected to think in code, and to get the feel of hands-on experience as soon as possible; programs are frequently developed by incrementally modifying other programs, and understanding is gained by making 'fruitful mistakes', so code that does not solve the problem is seen not as an error but as a useful step on the path.

Secondly, there is considerable social pressure to conform to the local culture. Where Pascal is fashionable, it is common to sneer publicly at Basic ('encourages hacking, rots the brain') and Fortran (a 'dinosaur'). Pascal programs written in the style of Basic, with many 'go to' commands and with no proper use of procedure structure, incur real opprobrium. But Prolog experts sneer at programs making too much use of cut-and-fail as 'Pascal written in Prolog', asserting that programs should be conceived declaratively not procedurally. And so on for other languages. Adherents of one language frequently claim that learning another language first will ruin the learner's chances of ever being able to program 'properly'.

Thirdly, in certain circumstances there are clear demands for a particular culture. Development laboratories need freedom to experiment; but in commercial software production of large systems that are intended to have a long life, each program has to interact with other programs, written and maintained by other people, over long periods of time, and the impetus must be towards standardization and simplification – similar solutions to similar problems, standard coding and documentation styles, and formalization of change procedures. The non-conformist, either a would-be standardizer in a development laboratory or a carefree explorer in a production team, would experience very strong pressure to change.

While differences in tradition are inevitable, and in any case are probably quite healthy, the arguments would often be assisted by some proper data. Can any pedagogic advantage be demonstrated for any of the different traditions? Is it really true that choice of first programming language seriously affects subsequent programming ability? Yet again, the evidence is lacking.

3 The environment

Much of the work described in this book deals with writing and reading programs using the simplest of technology, just pen and paper or a simple text editor. In truth, that is how most programming is still done. Yet computer scientists have been remarkably fecund in their creation of alternative programming environments, and many interesting psychological problems arise. The design of an environment presumably reflects someone's view of how to simplify some important types of programming task, but unfortunately the reports do not usually go into much detail about what tasks are intended to be supported. Furthermore, virtually no research has been reported on the advantages of different programming environments, which means that yet again we can do little beyond point out some of the questions that come up.

3.1 Neats and scruffies again

The characteristics of neat and scruffy environments reappear, of course, in their surrounding environments. The neat environment is closed and well defined; it is designed to encourage methodical, well-documented working, in which the coding stage is not started until the program is well-designed. Just as a single document completely defines the language, so the environment is well defined, with much emphasis on standardization across different locations.

The scruffy environment supports 'evolutionary' construction: users can slap a bit on here and a bit on there as their ideas develop. Xerox's Interlisp is a superb example. Not only can programs be built incrementally but Interlisp's 'advise' functions also allow second thoughts to be stuck in without having to recompile or even understand the original programs. Even the system programs can be modified by advise functions, allowing the environment itself to be modified without having to recompile all the source code. There is a tool for making programs try to run even in the presence of programming errors; this tool, called DWIM ('Do What I mean') completely defies the neat doctrine that a single document fully defines the language, because the behaviour of the program now depends entirely on how DWIM interprets it. The environment is open and extensible, so that from month to month new bits appear. The technology of electronic mail and of networks allows the new versions of the environment to be created frequently and mailed to users. Additions to the library of packages can easily be made by any competent user. Discovering what facilities are actually available in all these shifting sands can be quite difficult for newcomers: the simplest way to find something is to ask an expert. Smalltalk-80 is another extensible environment, which can be manipulated by programmers to suit their own needs or preferences. Expert advice on the available resources is needed very often by baffled newcomers to Smalltalk, just as in Interlisp.

3.2 The question of level

Programming languages (even low-level ones) contain a great deal of structure. It is possible to build and manipulate programs using high-level operations that reflect this structure. There are many differing opinions about high-level programming environments, reflecting differing beliefs about how programmers do, might or should think about programs and their construction, interpretation and manipulation.

Text-based editors remain the favourite choice, despite all subsequent developments. Present-day editors for use in programming environments provide some help with formatting programs, and editors specialized for programming use provide help with particular awkwardnesses – for example, editors for use in Lisp, with its many parentheses, provide some means to indicate which parentheses are paired together. A few editors, notably Emacs, can be customized to provide templates for lengthy and frequently used constructions, such as loops. In other respects the text-based

To start writing a function the user selects DEFINITION from the menu, which automatically generates the following code:

```
(DEFUN <NAME> (<PARAMS ...>)
       <FN-BODY ...>)
```

The angle brackets, <...>, surrounds slots that must be filled appropriately. Suppose the user next types the name FACTORIAL, giving:

```
(DEFUN FACTORIAL (<PARAMS ...  >)
       <FN-BODY ...>)
```

Next the user chooses to insert an IF-construct into the <FN-BODY> slot:

```
(DEFUN FACTORIAL (<PARAMS ...  >)
       (IF     <IF-TEST>
               <THEN-CASE>
       <ELSE-CASE>
       <FN-BODY ...>)
```

This goes on until the function is complete.

Figure 2: How Struedi, a Lisp editor, is used.

editors have changed little. Their users like them partly because they are simple to understand and do not add much of an extra learning load. More importantly, they do not constrain their users in any way at all. Programs can be built in any order – top-down, bottom-up, middle-out; pieces of code can be moved around the design at will; code can give way to remarks like 'I must finish this bit later'.

The major alternative is the editor based on syntactic structure, allowing users to manipulate their programs at the level of statements or expressions within statements. A fairly typical example among many others is Struedi (Köhne and Weber, 1987; Figure 2), designed to help novices learn to use Lisp without being bothered by syntax problems. It is based around templates of the language constructs, which can be filled in from a menu of various possibilities.

Although its design philosophy is not unusual, Struedi was built with exceptional care for human factors, and its authors even report a small experiment which suggested that Struedi helped with semantics learning, presumably by eliminating overheads spent on the syntax problems of Lisp (for a similar result, see Sime *et al.*, 1977). This sort of result is very encouraging to those who believe that learners suffer unnecessarily at the hands of badly designed languages presented in environments that give them little help.

Struedi is a neat language. It imposes an order of development on the programmer, and it insists that all changes to the program are made by modifying the source code, so that there is one and only one text defining the current state of the program. The code cannot be run until all errors have been removed from the text, so what the text says is a true and faithful indication of the program's behaviour. The editor is based on a 'parse tree', a syntactic analysis of the text, which means that like several other systems mentioned here, it is based on control flow. It is debatable whether that is the information the programmer most needs.

A different approach was taken with 'KBEmacs' (Waters, 1985), a knowledgeable program editor in which the user can call on a library of programming clichés to help construct the program. A *cliché* is a typical program fragment, such as 'file enumeration', and the user can call it up by asking for it by name. However, the cliché contains more than code: it also specifies *roles*, such as the input role defining the file or other structure, the *empty_test* which determines when to stop, the *element accessor* which accesses the current element in the structure, etc. In the example shown by Waters (p.1308), the user issues a short sequence of commands, starting:

```
Define a simple_report procedure UNIT_REPAIR_REPORT.
Fill the enumerator with a chain_enumeration of UNITS and REPAIRS.
etc
```

A total of six such lines is sufficient to generate more than fifty lines of Ada, forming a complete and correct procedure definition.

KBEmacs is not meant to be used as a novice's tool, but as an aid to expert programmers. The intention is to formalize their knowledge so that it can be used by the computer. Every expert programmer is expected to know that a 'simple report' cliché will have a role which enumerates elements, and to be acquainted with the cliché of 'chain enumeration'. All the programmer has to learn is the vocabulary – what the clichés and roles are called. KBEmacs preserves the cliché structure while it builds code, so that subsequent editing can refer to roles ('Fill the enumerator with <something else>').

Waters makes two points that are important for an assessment of KBEmacs as a partner to humans. First, unlike many systems it does not force the user to preserve syntactic correctness, nor does it impose a pure top-down development. According to Waters (p. 1318) it 'supports an evolutionary model of programming wherein key decisions about what a program is to do can be deferred to a later time as opposed to a model where only lower level implementation decisions can be deferred'. Secondly, clichés operate in the domain of programming plans, rather than of text or parse trees, so it is fundamentally unlike the systems mentioned above.

Unfortunately, Waters adds, attempting to externalize programming knowledge would be a 'lengthy' undertaking, since programmers probably know thousands of clichés, whereas the 1985 demonstration version of KBEmacs knew only a few dozen and was already some 40k lines of Lisp code. It was also 'fraught with bugs' and ran very slowly. Nevertheless it is an exciting indicator of possible lines of progress.

3.3 What to display

In Chapter 2.2 ('Programming Languages as Information Structures') I review the main findings on the problems of visual representations of programs. As will be seen, many questions are left unanswered by present research. One of the major questions is just what should be displayed, given today's powerful screen-based environments. Some visual programming systems go little further than to translate a standard

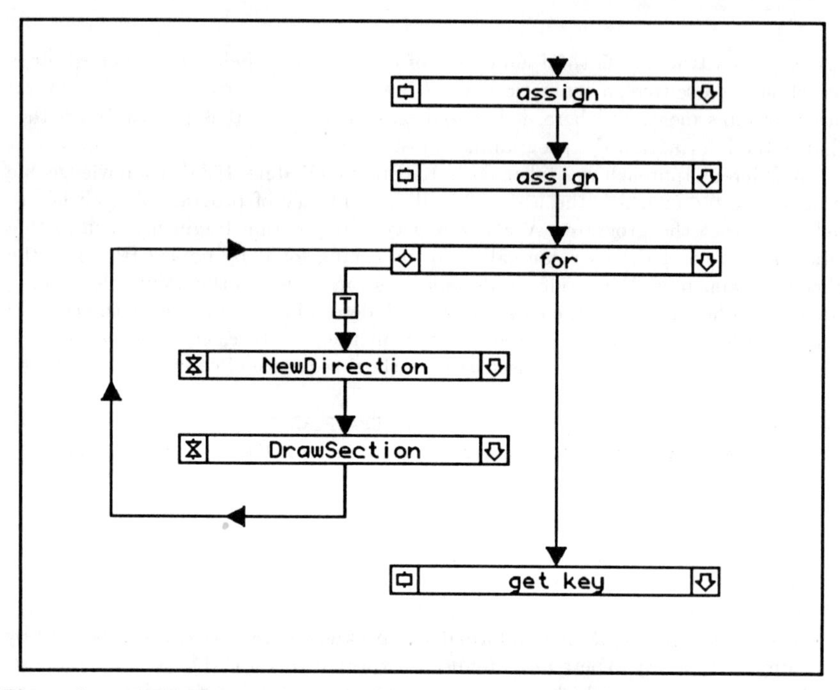

Figure 3: The VIP (Visual Interactive Programming) system, by Mainstasy Inc., provides an automated flowchart. A menu of tools allows the programmer to construct and manipulate the diagrams. Each box can be 'opened' to give more detail: the 'for' box, for instance, contains information about the details of an iterative loop.

programming language into a visual representation. A good example is Mainstay's VIP (Visual Interactive Programming), which presents the programming constructs of Pascal as labelled boxes, with arrows to mark the sequence of operation. A menu of tools provides for adding to and manipulating the diagram. VIP, as Figure 3 shows, is little more than an automated flowchart; this is evident in the level of abstraction, the vocabulary of 'assign', 'for', etc, and the choice of 'sequence of operation' as the main type of information to represent. Criticisms of structured flowcharts, such as those summarized in Chapter 2.2, include poor modularity and rigid control structures (see Green (1982) for a full account of research findings to that date). Such criticisms will probably apply to VIP with little change.

Other systems present information that is harder to represent in text. For example, good textual conventions for indicating parallel or concurrent execution are hard to devise: however hard we try to ignore the essentially linear nature of text, the mere fact that line A precedes line B insidiously murmurs to us that A is executed first. Putting special parentheses around the two lines to indicate that 'There is no constraint on order of execution within these bounds' is not an adequate solution. An example of a purely textual language designed to represent concurrent processes in that way is Occam, which uses the keywords `SEQ` and `PAR` to indicate sequential and parallel execution.

This Occam fragment shows one possible control structure for asking for two quantities to be input and then outputting the result of a computation:

```
SEQ
  PAR
    Width  := askWidth
    Length := askLength
  printArea (Width, Length))
  beep
```

The SEQ directs the computer to perform the subsequent commands in the sequence specified, that is, the paragraph starting with PAR (within which the two commands can be performed in any convenient order), then 'printArea', then 'beep'. Occam cannot show that the nature of the constraints is different for printArea (which logically cannot be executed until its data is ready) and for beep (which is executable at any time, but which the programmer has chosen to have executed only when all else is complete).

Parallel or concurrent execution is nevertheless relatively easy to represent diagrammatically. There is no built-in linear ordering in a diagram, it can simply be decreed that every box on the page is allowed to commence execution whenever the scheduler sees fit. 'Prograph', by The Gunakara Sun Systems, is such a language. The connecting lines in Prograph diagrams show *data flow*, not order of execution, and each box can be executed as soon as all its data is available. Different box shapes indicate different types of box: some supply constants as data, some take input from other sources, some invoke primitive operations, etc.

Programs are organized as 'methods', and the display is organized in terms of windows, each window representing a method. Figure 4 shows a simple method, using box shapes equivalent to subroutine calls – i.e., they denote methods defined in other windows. Because Prograph is a concurrent language, there is no constraint in this program on which of the top two boxes is executed first, 'ask width' or 'ask length'. However, 'print area' cannot be executed until both those have been executed, because it takes data from them. Thus Prograph's notation seems superior to Occam's in these respects.

But, given a visual display, the central question is *what to display*. VIP presents a simple translation of standard text into graphic form, as many other systems have done; for evaluations, see Chapter 2.2. Prograph presents material that would be hard to display as text in a simple form. Likewise, Brayshaw and Eisenstadt's 'Transparent Prolog Machine' and Böcker and co-workers' 'Kaestle' system try to *complement* the text: TPM animates the execution of Prolog programs, which is hard to deduce from the Prolog text, and Kaestle displays animated views of data structures in Lisp programs, likewise hard to deduce from the text (Brayshaw and Eisenstadt, 1988; Böcker *et al.*, 1986).

3.4 Browsing through programs

As programs get larger, programmers find it harder to locate the information they need. The techniques that have been developed to help them 'browse' include a wide range of very straightforward techniques, answering direct questions like 'where

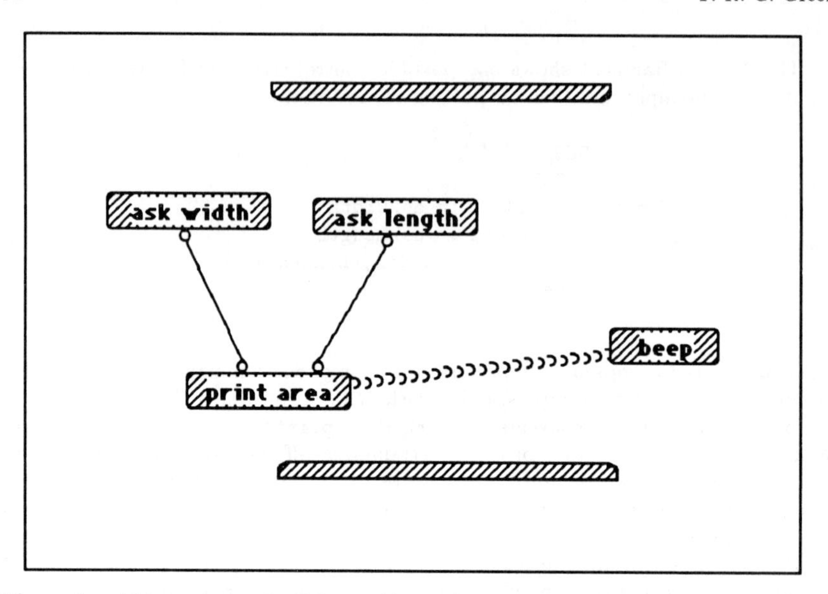

Figure 4: This is one 'method' from a Prograph program, itself calling further methods. Solid lines show data flow (from top to bottom); the 'horseshoe' line shows a user-defined constraint that 'print area' must be executed before 'beep'. In the absence of constraints any method may be executed as soon as its data, if any, is ready; thus this program does not specify which of 'ask width' and 'ask length' is to be performed first.

is this identifier used?' There are also some more interesting techniques based on psychological conjectures, either about the search patterns of people browsing large programs or else about the way information is mentally represented.

In the first category come browsing methods built on 'hypertext'. The conjecture is that with conventional systems programmers frequently need to interrupt their search for information about one aspect of their program to find out what something else does. Keeping track of where they were, and finding their way back, becomes a serious overhead. Hypertext systems provide built-in links, so that by say pointing to an identifier and querying it, the programmer is immediately shown the definition of the associated subroutine or data structure. Experience shows that hypertext users lose track of where they came from, so good systems help the programmer keep track of the search path and find the way back; for example, the recent 'Hybrow' system (Seabrook and Shneiderman, 1989) provides special support to help in managing the many information windows that make up the search trail.

Quite different in spirit are browsers that try to filter out irrelevant information. Among many such, the 'fish-eye view' (Furnas, 1986) deserves mention. Instead of showing the programmer a simple slab of C text, as much as will fit on the display, the fish-eye view in his example lists only the lines that seem to be important in providing context around the line that is currently being edited. Another example from that paper is a 'fish-eye calendar' where the current day is shown in detail, the current week in less detail, and the current month in still less detail. A similar approach, less well worked out, is taken in certain structure-based editors.

Browsers like these are obviously built in response to conjectures about what information programmers want and how they would like to receive it. There has been very little serious work on that issue. The simple conjecture of the previous section is that the greatest effort should go into making it easy for programmers to get at the information that is not easily extracted from the text, which implies that the fish-eye view might give its users less help than the hypertext browser. But that is yet another unresolved issue.

4 Sidestreams

So far we have concentrated on examples taken from what I see as the main line of progress in programming languages and environments. Around this line have developed huge numbers of experimental or special-purpose languages and environments, driven by the needs of the moment or by a spirit of exploration, and some interesting experiments in combining paradigms that appear at first sight to be incompatible. No survey would be complete without mentioning at least a few of these and considering whether they raise special problems, outside those raised by the more conventional languages.

4.1 Specialist languages

Some unusual and interesting designs for languages have developed from the need to control specific pieces of equipment. One of the earliest was Forth (Brodie and Forth, Inc., 1987), developed originally to control astronomical observation equipment. Since the control equipment was driven by primitive microprocessors, the overheads of conventional high-level languages would have been unacceptable, and in any case it was necessary to stay very close to the machine level. Nevertheless pure assembly code was unworkable – it took too long to write and debug a program, and changing the program was too hard. Forth was a compromise between the virtues of assembly code and the virtues of high-level languages. It has become a popular language for several types of work, including robotics, and has continued to develop – there has even been an object-oriented Forth. Recently the effectiveness of Forth as a compromise between high and low levels was underlined by the release of PostScript (Adobe Systems Inc., 1985), a device-independent page-description language now widely used to control laser printers. PostScript is comprehensible to humans and can be used as a programming language, yet it also serves as a communication medium between computers: word processors and graphics programs generate PostScript programs which are executed by the microprocessor in the laser writer or other output device.

The main idea behind Forth, continued in modified form in PostScript, is thoroughgoing submission to the reverse Polish notation (RPN). In RPN, operations are expressed as $AB + C *$ rather than as $(A + B) * C$. Certain types of hand calculator do much the same. The difficulty about RPN is how to express control structure, i.e. loops and conditionals, and how to cope with arrays. Forth boldly adopts an RPN solution to all these. For instance, the following code defines a 'word' (subroutine) that will print a warning message if its argument is greater than 220:

```
: ?TOO-HOT  ( temperature -- )
    1  >  IF  ." Danger -- Reduce Heat!  " THEN ;
```

Notice the terse syntax: : is the keyword that introduces the definition, ." prints the following symbols up to the closing quotation marks. Notice also the RPN conditional: the keyword, IF, follows the predicate, and the closing keyword THEN follows the action arm. The Pascal version would of course be:

```
IF temp > 220 THEN writeln('Danger ..').
```

But the most striking aspect of Forth is the free use of the execution stack. The programmer is at liberty to leave values on the stack for as long as desired. When ?TOO-HOT is executed it expects to find a value on the stack. In most high-level languages, a formal parameter would be declared and on entry to the subroutine an automatically compiled operation would remove the parameter from the stack and put it into the parameter variable. Forth leaves all that to the programmer: less complexity in the system, faster speed of execution, but more risk of error. 'Scruffy heaven' is its philosophy on all counts. Questions raised by Forth are: Is it readable? Can we parse for structure? Is it error-prone? Is it easy to learn because of having fewer new concepts?

Forth might look a bit old-fashioned today, but it must be understood that a whole heap of historical, organizational and economic factors determine choices of industrial languages. The same is true of languages for numerical control and for programming robots. Visser (1988) describes several types of language in use for computer control of automatic tools, one of which (MODICON) is based on relay closures, represented either as schematic relays or as Boolean expressions, with intermediate variables holding subexpressions. Serious and lengthy programs are developed in this language.

Now for a very different group of specialist languages, developed as sequence controllers. Sequence control languages typically contain no conditional structures but they provide repetition of patterns and of groups of patterns, and some means to define new patterns. Languages of this type have been developed for use in diverse areas. Advanced knitting machines aimed at the domestic market are programmed using pattern-description languages capable of creating a wide range of effects, yet intended for use by people with no background in informatics. In contrast, sequence controllers for musical effects (synthesizers, drum machines, tape editing for video, etc.) are likely to be used in a 'hi-tech' background. Many of these controllers are meant for professional use, but a few years ago an inexpensive drum machine for a domestic computer (the 'SpecDrum', distributed by Cheetah Marketing Ltd, for the Sinclair Spectrum) already provided remarkable capability. The user could store sixteen 'songs', each consisting of up to 255 sections, each of which could be repeated up to 255 times. In each section, one of a list of patterns was performed, a pattern consisting of some number of bars of however many beats was convenient. Six different types of drum sounds could be inserted into the bars, both on and off beat, and the tempo of the song could if required be over-ridden by an individual pattern.

Sequence control looks simple, but here are new questions for the list. First, what degree of abstraction is tolerable – could the SpecDrum user have coped with

Language	Example Query for 'Find the names of employees in department 50'
SQL:	SELECT NAME FROM EMP WHERE DEPTNO = 50
QBE:	

EMP	NAME	DEPTNO	SALARY
	p. Brown	50	

Figure 5: Two designs for query languages (from Reisner, 1988). Although much simpler than programming languages, these languages share some of the characteristics. The advantages and disadvantages of different designs are still not well understood.

subpatterns, and subsubpatterns? Second, if the system is to be readily usable, must the degree of abstraction be predetermined (songs, sections, patterns), or can it be extended by the user at will? (Remember that if the degree of abstraction is to be extensible, the user needs not only a method to define and call patterns, but also a method to define new levels of subpatterns...). Achieving such complexity of structure without making the system impossible to use in a domestic setting will tax the designers!

The last examples of specialist languages that I shall mention are query languages for database searching. Query languages are somewhat outside the major scope of this book, yet ultimately when we understand how to design programming languages and other notations we should be able to deal with query languages on the same basis. Two well-known query languages are SQL and QBE: Figure 5 (from Reisner, 1988) illustrates the same query in each representation. Ideally, we should be able to predict the pros and cons for each design. Unlike most other areas, there have been controlled experiments on query languages, very well reviewed by Reisner (1988): yet she concludes (p. 267), 'It is clear that at this stage of our knowledge only a few of the issues have even been identified, much less studied. Clear guidelines to aid good design do not exist'.

4.2 Mixed paradigms

Many problems seem to be peculiarly intractable to any single programming paradigm. If approached procedurally, it becomes clear that part of the problem is best approached declaratively, and vice versa. Why not use a mixed paradigm, and treat each aspect of the problem on its merits?

An interesting early development was a scheme for transforming a typical procedural program to satisfy a number of 'sequence relations' (Middleton, 1980). Examples of the type of relation covered are: 'whenever the number of times that A has been executed is a multiple of N, do B'; 'stop as soon as the number of times C has been executed is M' (where the action C might occur in more than one location in the original program); 'between actions D and E, do F at least once' (e.g. between opening a file and terminating, make sure it has been closed – whatever path is taken). So a complete program could have two parts, a standard Pascal-like part plus a declarative part specifying sequence constraints, and the compiler would automatically combine them. Middleton points out that meeting these and other requirements in conventional sequential terms causes a huge proliferation of administrative variables, keeping track of how often various events have been performed. A toy example shows a very simple seven-line program, comprising a loop and a conditional, being transformed to meet four sequence relations; the final program is thirty-four lines of vicious-looking code, and it seems undeniable that the mixed scheme is more comprehensible and easier to modify.

The Middleton scheme has not been tried as a working system, but another approach has: the combination of Pop-11 (a procedural language with Lisp-like data structures but a more expressive syntax) and Prolog, to form PopLog. In principle, a recalcitrant problem can be solved partly in Pop-11 and partly in Prolog, with data communicated between the two systems. Although this scheme has been available for some years now, as have similar schemes based on Lisp rather than Pop-11, this combination has made little impact outside its own band of aficionados. The reasons are not clear, for it would appear to yield precisely the information that the user wants in the most digestible form.

5 Where next?

We shall all be programmers soon. That, at least, is the impression one receives from the vast spread of programming possibilities, creating new environments such as the knowledge-based and visual systems illustrated above; new models of programming such as logic-based and constraint-based programming; and new applications, to science, learning, and education, to the office world, the domestic world, and the leisure world. Some idea of where programming is going can be got by looking at a few recent developments. This cannot possibly be representative; for instance, I am specifically excluding all educational programming systems, which will receive a brief treatment in a later chapter, and making no mention of program generators, fourth-generation languages, very high-level application-specific languages, or the prospects in languages for embedded systems such as domestic microprocessors. My aim in this closing section is to pick out what may well be seminal developments in programming styles.

5.1 Spreadsheets

The spreadsheet is now a familiar computational tool. It is so simple to use that in many people's eyes it hardly counts as programming, but merely as a declarative statement of relationships between numbers and formulae. The computation of area (used to illustrate Prograph above) can be trivially expressed in a spreadsheet, by putting the two values in two cells and putting the multiplication formula in a third cell. Other domains are now sometimes packaged with a spreadsheet-like interface, and this style may become even more widespread. For instance, Spenke and Beilken (1989) describe Perplex, which interestingly 'combines the power of logic programming with the popular user interface of spreadsheets'; the user creates predicates by successively refining first-solution attempts, with immediate knowledge of results, as with conventional spreadsheets. 'There is no new formalism or language the user has to learn in order to define new predicates. Programming general solutions is almost as easy as solving a single, concrete problem. The user need not even know in advance that he is writing a program.'

5.2 Knowledge engineering

'Knowledge engineering' describes the explicit computer-based representation of human knowledge of how to perform skilled tasks, such as bidding in contract bridge, choosing suitable crops for farmers, or identifying the cause of an emergency in a nuclear reactor; most commonly such knowledge is represented as an 'expert system', which contains the knowledge used by experts and can be consulted for advice in place of an expert. Vast claims have been made for them, giving the impression that can readily be built by persons with no specialist programming experience, that real expertise can be captured, and that they can be used by clients with very little expertise as a genuine alternative to consulting a live expert. In actual fact, they are at present useful but quite limited. They will probably become more common and more useful in the next decade, but despite the early claims it is unlikely that systems of any complexity will be built by non-programmers or easily used by non-specialists.

The area has been dominated by representational techniques using 'production rule' systems, programming languages built around the IF-THEN construct. A commercially available example is 'Xi Plus' which uses a relatively comprehensible, English-like syntax, in the form:

```
if temperature < 55
then room is cool
```

A set of such rules, plus a set of facts established for the current case (e.g. temperature = 51), forms a 'knowledge base'. Figure 6 gives as an example a fragment of a knowledge base for choosing house plants for a given room. The rules can be used in two ways – either by supplying some facts about the room in question, and then asking Xi to report what conclusions follow ('forward chaining') or else by asking specifically to find out whether a given conclusion is true ('backward chaining'). In forward chaining the expert system examines all its rules: any rule whose IF part is satisfied by the existing facts can be executed, and the statements following the THEN part are added to the list of current facts. In backward chaining, the expert system first examines rules whose THEN part can establish the desired goal: if these rules have an IF part which only uses currently known facts, well and good; otherwise, the system tries to find a rule whose THEN part can satisfy that IF, and so on. These two schemes produce very different behaviour, and Xi Plus is unusual in allowing not only both modes, but also a smooth transition from one to the other depending on circumstances.

Expert systems, even more than other forms of programming, aim to put expert knowledge into a tractable and comprehensible form. One favoured technique is to allow the client who uses them to ask 'why?' questions: 'Why is ivy the best plant?' receives the reply 'Because the light is poor and the room is cool'. Essentially, this is a form of directed browsing over the execution history of the program. Whether this

```
if temperature < 55
then room is cool

if temperature >= 55
   and temperature < 65
then room is warm

if temperature >= 65
then room is hot

if light is poor
   and room is cool
then best plant is ivy

...........................
...........................

if light is sunny
   and room is hot
then best plant is collection of cacti

question light is sunny, bright, poor
text How good is the light in your room?

question temperature = 45 to 75
text What is the temperature of your room in
degrees Fahrenheit?

query best plant
```

Figure 6: Fragment of an expert system built in Xi plus. The IF-THEN rules form a knowledge base. In response to the instruction 'query best plant', the system looks for rules that can establish a best plant. On finding rule 4, which applies in poor light, it looks for facts or rules about light; the instruction 'question light is ...' tells it to ask the user 'How good is the light in your room?' and to use the reply as a newly established fact. The reply 'poor' will allow it to go on to '...and room is cool'; other replies will cause it to discard the ivy rule and look for another possibility (not shown here).

```
IF
   When:
      Every day at:  5:00 AM
   Header contains:
   Msg body contains:  <string>
   Folder name is :  In-Tray
   Msg length is:
   Msg is:  Read AND NOT Tagged
   Date sent is:
      More than 2 days ago
THEN
   Move-to Old-Mail
```

```
WHEN daily(5:00 AM)
   FOREACH msg IN "In-Tray"
   IF read(msg) AND
      NOT (tagged(msg) AND
         date-sent(msg) > 2 days ago
THEN
   move-to-folder(msg,"Old-Mail")
```

Figure 7: Example of forms-based version (left) and language-based version (right) of the same program (from Jeffries and Rosenberg, 1987). The task is at 5 a.m. to take all the old messages (that is, ones that have been read and have not been tagged for keeping) out of the 'in-tray' and put them into an 'old-mail' folder.

is the best way to help the client understand how the decision was reached remains an open question.

Also at issue is whether such systems will really become favoured vehicles for non-programming specialists to communicate expertise to others. One of the few studies of difficulty in this type of system (Ward and Sleeman, 1988) reports a wide variety of minor difficulties with the notational details, suggesting that the style of language has to evolve further. More importantly, the programmers were observed to meet problems requiring 'deep and careful thought', especially to do with such issues as 'How should rule premises and rule conclusions be organized, and how should rules inter-relate?'

5.3 Forms programming

Another application of the IF-THEN format is in 'programming by form filling', which is intended to provide a simple method for non-programmers to build programs in limited domains. The limited possibilities of the domain can be exploited to replace conventional procedural language style with multichoice techniques. The example (Figure 7) concerns a programmable handler for electronic mail.

Form-filling methods give users fewer ways to go wrong than conventional languages, so it is not surprising that non-programmers do better; more surprising, however, is the fact that experienced programmers also scored slightly better with the form-filling methods. The authors point out that these systems are 'close to the prototypical tasks the user wants to perform', so that the translation distance between the goal and the required actions is considerably reduced. This is an interesting explanation, although there has been no report of a test such as presenting the two versions of Figure 7 to users and asking which one is 'closer to the task'. The sceptic might suggest that one version does a better job of reminding users about what conditions to include and how the syntax works! More use should, perhaps, be made of techniques which have such a property.

5.4 Everyday programming

The most challenging task is to get everything right at once: a programming language that is easy for beginners, has enough power for experts, comes with an environment which meets the user's needs, and is attractive to use. While we are a long way from achieving that, some interesting possibilities are emerging, among them HyperCard for the Macintosh. HyperCard reverses the normal order of things, in which the programmer writes a program which might, rather painfully, put some graphics on the screen. Instead, HyperCard provides ample tools for users to build graphics as in a normal painting-style application, and then allows fragments of program to be attached to the resulting picture. The fragments are usually attached to 'buttons' which, when pressed, cause them to be executed. One single screen is called a 'card', and a HyperCard program in fact consists of any number of cards, each of which can show a different picture. A simple example is shown in Figure 8; the user is expected to use the mouse and keyboard to write numbers into the two fields, and then to 'press' the button to cause the computation to be performed.

A limited form of inheritance caters for the frequent need to produce a set of cards with the same basic design, such as all having a button in the same place. All cards have a 'background', which can be shared by any number of cards; and so a button which is placed on the background will be present on each card. Both scripts and graphics can be inherited from the background, and further levels of inheritance are also provided for scripts.

Apple, the creators of HyperCard, clearly intended it to become as near an everyday programming language as they could manage, and it is instructive to consider some of the steps they took. First, they tried to give it wide *everyday applicability*, not so much by building in computing power as by providing means for HyperCard to control domestic gadgetry. With the right attachments, it can control compact disc players and video recorders, synthesize sounds, run a MIDI interface to audio equipment, and even dial telephone numbers; in the office it can provide calendars, reminders and organizers. Next, the system is meant to be *foolproof*. There are no variations depending on the particular hardware, the system has been very thoroughly debugged, it is hard (but not quite impossible!) to create catastrophic errors which delete important material, and the error reporting is almost jargon free. They also emphasized *simplicity*. Difficult concepts have been avoided. (Most data structuring methods have been classified, rightly or wrongly, as 'difficult', so there is no form of linked list and only a rudimentary form of array.) The programming language has been kept free of special symbols and funny-looking words, and although the result is rather verbose at least the user can usually remember what the syntax rules are. The use of inheritance techniques is intended to *reduce drudgery*, although this has not been entirely successful – more powerful tools are needed. Finally, and perhaps most innovative, is the principle that *program fragments are attached to screen objects*, not the other way about.

HyperCard has the faults of its virtues. The language has little expressive power, and it is verbose and tiresome to proficient programmers. By attaching program frag-

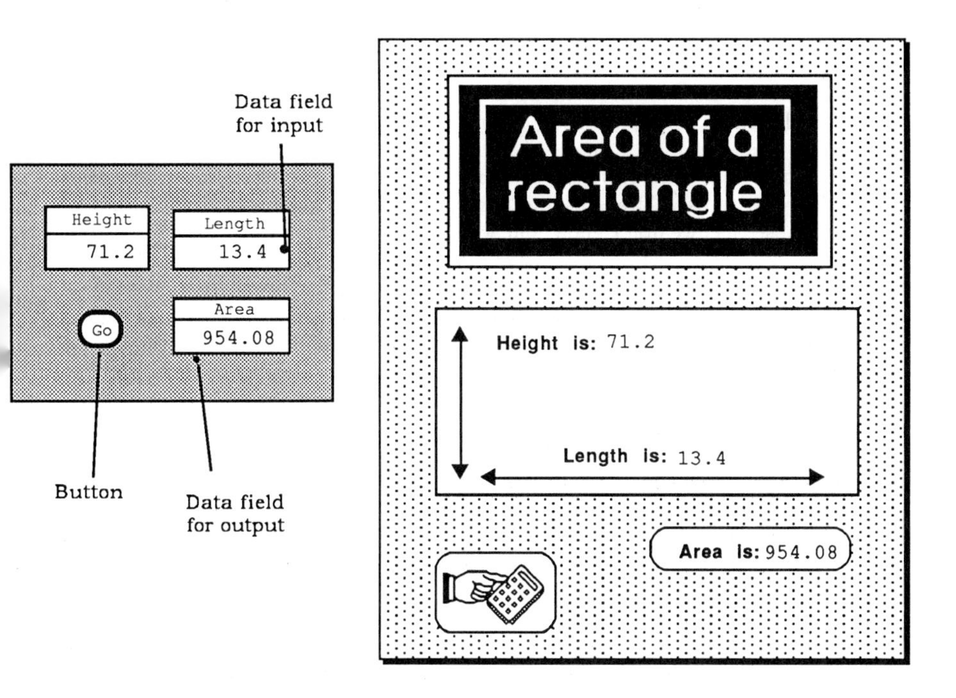

Figure 8: HyperCard programs are attached to graphic objects, typically buttons, on 'cards'. To the user they look like active diagrams. Here are two different possible visual arrangements for the same simple example, in which users enter data in 'fields' and press a 'button' to start the computation. In this example, the 'script' (program) for the button 'Go' might be:

```
   on mouseUp
        put card field "length" * card field "height" into card field
"answer"
   end mouseUp
```

Scripts are executed when a given event occurs. This script will be executed when the mouse button is released ('mouseUp') within the screen area of the button. Neither the script of the button nor the names of the fields are visible in the normal mode: the programmer has to 'look behind' them. More complex programs may relate several different cards: e.g. successive cards may handle area computations for a variety of different shapes.

ments to screen objects the overall visibility of the program has been very seriously reduced (you can't easily see the scripts for two buttons at the same time) and programmers have to commit themselves too early to decisions they may later revise. It will be interesting to see what the next attempt looks like.

6 Conclusions

The development of computing has been based largely on guesses about what people would find easy to use. After this quick tour, what do we know – or better, what do we now realise that we don't know, but would like to?

(1) We still think too readily of programs as just being for compilation. We should think of them also as being for communication from ourselves to others, and as vehicles for expressing our own thoughts to ourselves. So we should think more about reading versus writing, capture of ideas versus display of ideas, etc.

(2) Existing environments are better at displaying program information than at providing means to manipulate it effectively. We need more research on manipulation of complex structures.

(3) Not enough is known about the advantages of the techniques currently being explored, such as program animation, mixed paradigms, etc. Surely these deserve active investigation.

(4) There has not been enough conscious effort to find ways to display alternative representations of information or structure; instead, many systems simply translate textual information into an equivalent form.

(5) Where new types of information structure have been devised (e.g. representations based on programmer's knowledge, such as KBEmacs), there is often not enough existing psychological research to be useful.

(6) The relationship between the structures of programming languages and the tools for operating on programs has not been well explored. Research on each tends to occur without considering the other. Should we not try to discover, at a generic level, what support programmers need (or at least, programmers of a particular experience level), and then when designing tools and languages try to ensure that nothing is omitted?

(7) Is there a 'solution' to such profound differences as the argument between the 'neats' and the 'scruffies'? Probably not. Perhaps we should aim for an agreement to co-exist, with different approaches suiting different personalities of different contexts. This demands more research into individual differences – not merely in aptitude.

In short, we know something about the psychology of learning to program, of understanding programs, etc., but only in a very restricted range of languages and environments. It is important to extend our studies to wider vistas.

Acknowledgements

I should like to thank Heather Stark, Barbara Kitchenham and Marian Petre for their helpful comments on an earlier version of this chapter.

References

Adobe Systems Inc. (1985). *PostScript Language Reference Manual.* Reading, MA: Addison-Wesley.

Böcker, H. D., Fischer, G. and Nieper, H. (1986). The enhancement of understanding through visual representations. *Proceedings CHI'86 Conference on Computer-Human Interaction.* New York: ACM.

Brayshaw, M. and Eisenstadt, M. (1988). Adding data and procedure abstraction to the Transparent Prolog Machine (TPM). *In* R. A. Kowalski and K. A. Bowen (Eds), *Logic Programming: Proceedings of the 5th International Conference and Symposium.* MIT Press.

Brodie, L. and Forth, Inc. (1987). *Starting Forth.* Englewood Cliffs: Prentice-Hall.

Détienne, F. (1990). Difficulties in designing with an object-oriented programming language. To be presented at *INTERACT '90 Conference on Computer-Human Factors.* Cambridge, England.

Furnas, G.W. (1986). Generalized fisheye views. *Proceedings CHI'86 Conference on Computer-Human Interaction.* New York: ACM.

Green, T.R.G. (1980). Programming as a cognitive activity. *In* H.T. Smith and T.R.G. Green (Eds), *Human Interaction with Computers.* London: Academic Press.

Green, T.R.G. (1982). Pictures of programs and other processes, or how to do things with lines. *Behaviour and Information Technology,* **1**, 3-36.

Jeffries, R. and Rosenberg, J. (1987). Comparing a form-based user interface for constructing a mail program. *In* J.M. Carroll and P.P. Tanner (Eds) 'CHI+GI 1987', *Proceedings ACM Conference on Human Factors in Computing Systems and Graphics Interface.* New York: ACM

Kernighan, B.W. and Plauger, P.J. (1974). *The Elements of Programming Style.* New-York: McGraw-Hill.

Köhne, A. and Weber, G. (1987). Struedi: a Lisp-structure editor for novice programmers. *In* H.-J. Bullinger and B. Shackel (Eds), *Human-Computer Interaction - INTERACT '87.* New York: Elsevier.

Middleton, A.G. (1980). A program transformation system for implementing sequence relationships. Technical Report 8001, Dept. of Computer Science, Memorial University of Newfoundland, St John's, Newfoundland, Canada.

Neal, L. R. (1987). User modelling for syntax-directed editors. *In* H.-J. Bullinger and B. Shackel (Eds), *Human-Computer Interaction - INTERACT '87.* New York: Elsevier.

Reisner, P. (1988). Query languages. *In* M. Helander (Ed.), *Handbook of Human-Computer Interaction.* New York: Elsevier.

Seabrook, R.H.C. and Shneiderman, B. (1989). The user interface in a hypertext, multi-window program browser. *Interacting with Computers,* **1**, 299-337.

Sime, M. E., Arblaster, A. T. and Green, T. R. G. (1977). Reducing errors in computer conditionals by prescribing a writing procedure. *International Journal of Man-Machine Studies,* **9**, 119-126.

Spenke, M. and Beilken, C. (1989). A spreadsheet interface for logic programming. *Proceedings CHI'86 Conference on Computer-Human Interaction.* New York: ACM.

Visser, W. (1988). *Langages de prommation dédiés: quelques exemples dans le domaine des automates programmables industriels.* Technical Report, INRIA, Rocquencourt, France.

Ward, R.D. and Sleeman, D. (1988). Learning to use the S.1 knowledge engineering tool. *Knowledge Engineering Review,* **2**, 4

Waters, R. C. (1985). The programmer's apprentice: a session with KBEmacs. *IEEE Transactions on Software Engineering,* **11**, 1296-1320.

Chapter 1.3

The Tasks of Programming

Nancy Pennington[1] and Beatrice Grabowski[2]

[1] *Psychology Department, Campus Box 345, University of Colorado, Boulder, Colorado 80309, USA*
[2] *Department of Anthropology, Rawles Hall, Indiana University, Bloomington, Indiana 47402, USA*

Abstract

Computer programming and other design tasks have often been characterized as a set of non-interacting subtasks. In principle, it may be possible to separate these subtasks, but in practice there are substantial interactions between them. We argue that this is a fundamental feature of programming deriving from the cognitive characteristics of the subtasks, the high uncertainty in programming environments, and the social nature of the environments in which complex software development takes place.

1 Introduction

A distinctive characteristic of computer programming derives from the variety of subtasks and types of specialized knowledge that are necessary to perform effectively. A skilled programmer must comprehend the problem to be solved by the program, design an algorithm to solve the problem, code the algorithm into a conventional

Psychology of Programming
ISBN 0-12-350772-3

programming language, test the program and make modifications in the program once it is completed. Success at these programming tasks requires knowledge of the external problem domain (e.g. statistics, finance, electronics, communications), knowledge of design strategies to develop and implement algorithms, knowledge of programming languages, knowledge of computer hardware features that affect software implementation, and knowledge of the manner in which the program will be used. In the present chapter, we provide an overview of computer programming in terms of programming subtasks, knowledge sources and the interrelations between these components.

Computer programming may be characterized 'as a whole' as a design task (Greeno and Simon, 1988). Examples of other design tasks include architecture, electrical circuit design, music composition, choreographing a dance, writing an essay or writing an instruction manual. There are several features that design tasks have in common. First, the goal of the designer is to arrange a collection of primitive elements in the design language in such a way as to achieve a particular set of goals. For computer programming, this involves piecing together a set of programming language instructions that will solve a specified problem. Secondly, two fundamental activities in design task domains are composition and comprehension. Composition is the development of a design and comprehension results in an understanding of a design. The essence of the composition task in programming is to map a description of *what* the program is to accomplish, in the language of real-world problem domains, into a detailed list of instructions to the computer designating exactly *how* to accomplish those goals in the programming language domain (Brooks, 1983). Comprehension of a program may be viewed as the reverse series of transformations from *how* to *what*. Thirdly, the composition and comprehension transformations are psychologically complex tasks because they entail multiple subtasks that draw on different knowledge domains and a variety of cognitive processes. For example, Brooks (1977) divides the programming task into subtasks of understanding the problem, method finding (planning) and coding. Other researchers include design, coding and maintenance subtasks. Multiple subtasks are also typical of other design tasks such as writing, for which planning, translating and reviewing have been suggested as distinct component processes (Hayes and Flower, 1980).

In Figure 1 we depict this characterization of design tasks in general, and of computer programming in particular. Basic programming subtasks of (1) understanding the problem, (2) design, (3) coding and (4) maintenance are shown. Basic processes of composition and comprehension are shown as processing tasks that cycle through the different subtasks. With each subtask we have associated certain mental products and knowledge domains, tentatively adopting Brooks' (1983) definition of the programming process – the serial mapping from one knowledge domain to another, beginning with the problem domain, through several intermediate knowledge domains, and ending with the programming language domain – as a useful framework within which to begin analysis.

The tidy separation of programming subtasks and representations shown in Figure 1 is, however, misleading as a description of software design and programming as it usually occurs for any moderately complicated programming project (see Chapters 3.3 and 4.2). The programming subtasks described above have multiple interconnections that make them difficult to separate in practice. This is also true of other complex design tasks such as architecture, planning, art and writing, in which con-

Programming subtasks	Basic processes	Knowledge domains	Mental representations	External representations
Understanding the problem	COMPOSITION ↓ COMPREHENSION ↑	Domain knowledge (e.g., statistics, banking)	Situation model	Requirements document Specifications document
Design		Design strategies Programming Algorithms and methods Design language	Solution model Plan representation	Design document
Coding		Programming language Programming conventions	Program representation	Code
Maintenance		All knowledge domains Debugging, testing strategies Frequent kinds of error	All representations	All documents

Figure 1: The tasks of programming.

ceptualization, design and implementation will influence each other in all directions (Rowe, 1987).

One way in which there are interactions between programming subtasks is that programmers rarely complete one subtask before beginning the next. Rather, the process is better described as repeated alternation among subtasks. Thus the programmer may continue to work on 'understanding the problem' in alternation with design, coding, and revision (Malhotra *et al.*, 1980; Chapter 3.3). A second feature of the programming task that makes a clean decomposition into subtasks unrealistic is that design not only takes place at different *levels* of abstraction (more than one level of detail), but it involves multiple qualitatively *different* abstractions. Kant and Newell (1985) call these different 'problem spaces'; Pennington (1987b) labels the idea 'multiple abstractions'. This refers to the idea that the design may be described in terms of its functional specifications, its procedural interrelations, its structural or transformational properties. Different abstractions (or problem spaces) may be most useful during different subtasks, yet each is needed throughout the entire programming process (see also Chapters 1.1, 1.2 and 2.2). A third and related feature of programming that connects subtasks is the interaction between knowledge domains (Barstow, 1985; Kant and Newell, 1984). For example, the computational algorithms will often depend on knowledge of problem-solving heuristics in the application domain such as electronics, geometry or physics. A fourth aspect that complicates the programming picture is related to the multiple and often messy environments in which programming takes place (see Green, Chapter 1.2). Some environments allow easy interaction between subtasks, some impose stricter separations. In addition, subtasks are most often distributed over members of the project team, adding hefty doses of communications and social problems to already difficult intellectual tasks (see Chapter 4.1).

In sum, programming is a complex cognitive and social task composed of a variety of interacting subtasks and involving several kinds of specialized knowledge. We will use Figure 1, in its bare analytical form to organize the following sections that elaborate each subtask, returning throughout and at the end to further discussion of the interrelations between component tasks.

2 Understanding the problem and problem representation

Programming problems are unique in that they usually involve solving a problem in another (application) problem domain, such as mathematics, accounting, electronics or physics, in addition to solving the program design problem. For example, a computer program may control the scheduling of a set of elevators (Guindon *et al.*, 1987). Decisions about how the elevators should work are part of solving the application problem.

It is usually expected that these application decisions will be set out in a requirements document listing the client's goals, or the requirements may be established through interaction with the client. One thing that is now clear, is that requirements documents and client's statements of goals are never complete. In some cases clients may not know all of their goals at the outset (Thomas and Carroll, 1979); in other cases there are assumptions of 'common' knowledge that will be brought to bear in combination with the written problem description (Krasner *et al.*, 1987); in other cases there will simply be omissions. Thus, one critical aspect of understanding the problem to be solved is that the programmer possesses knowledge of the application problem domain (see Chapter 2.3).

Even in the blandest of problems, some domain knowledge will be required. For example, in the text indexing problem used by Jeffries *et al.* (1981), the problem description stated that the program must produce an index listing page numbers of all instances of certain terms in the text. It requires knowledge of texts to realize that there may be instances of terms in the text that are hyphenated and therefore need special treatment. Curtis and his colleagues, in an extensive field study of programming teams (Krasner *et al.*, 1987), found that one of the most critical problems facing teams designing complex systems was that programmers often lacked critical domain knowledge that would allow them to detect incompletenesses in complex requirements documents or in initial discussions of system plans. Adelson and Soloway (1985) demonstrated that inexperience in a problem domain was associated with programmer failure to develop a 'global model' of the design. Thus domain knowledge is critical for providing a context for interpreting requirements, detecting incompleteness, constraining initial design solutions, and developing a global design model. However, even *experts* in a problem domain will not interpret system requirements identically, especially when they are stated in high-level terms (Thomas and Carroll, 1979).

A second critical feature of understanding the problem is the *form* of the problem representation. By problem representation, we mean a mental conception of the problem to be solved and/or an external characterization in the form of text or diagrams. It is well documented in the psychological literature that the representation of the problem that is constructed is an important determinant of the range of solutions that will be considered and is also an important source of problem-solving difficulty (see e.g. Hayes and Simon, 1977). One frequently cited example of this is a brain teaser called the mutilated checkerboard problem in which the task is to decide whether or not a checkerboard that has certain squares removed can be completely covered by rectangles the size of two adjoining squares. Although the problem is most readily represented in the spatial terms of checkerboards and rectangles, it is most easily solved when it is represented in terms of the *numbers* of squares of each colour that remain after mutilation. Similar results have been found in studies of design: both the surface form (e.g. spatial or temporal) of the problem and the structuring of requirements will influence problem representation and consequently the characteristics of problem solutions (Carroll *et al.*, 1980; Ratcliffe and Siddiqi, 1985).

For programming problems it is clear that understanding the problem will result in a mental conception of the problem and that there will also be external notes and documents. Questions of interest concern what form the problem representation does take or should take, and what kinds of information are explicit at this stage of analysis. One view of problem comprehension and representation stresses an external problem domain model that includes descriptions of the various objects, their properties and relations, their initial and final states, and the operations available for going from initial to final states (Brooks, 1983; Goldman *et al.*, 1977; Miller and Goldstein, 1977). This conception of the appropriate problem representation is analogous to the development of a 'situation model', a mental representation of the situation to which the to-be-solved problem refers (van Dijk and Kintsch, 1983). For example, Kintsch has found that in solving algebra word problems, a key to success is having constructed an accurate mental representation of the real-world situation described in the problem. People who did *not* do this often correctly solved the *wrong* problem (Cummins *et al.*, 1988). Similarly in programming, Pennington (1987a) has found that the best programmers constructed a mental representation of the real world problem domain. Some, but not all prescriptions for programming embody this concept. For example, the Jackson design methodology (Jackson, 1975, 1983) advocates careful description of various domain objects, their properties and relations, as a first step in designing a computer program.

It has also been suggested that expert programmers have 'problem categories'. That is, when presented with a design problem, the designer will determine its nature by associating it with previously known problems. For experts, it is thought that each of the problem categories is closely associated with a problem solution plan that applies to problems of that type, in contrast to novices who might categorize problems according to surface features such as application area. Thus, when the expert recognizes that a problem is of a certain type, he/she also knows how to begin solving it. In an empirical test of this idea, Weiser and Shertz (1983) found that novices tended to categorize problems by application area and experts classified by algorithm. Other studies of novice programmers suggest that the ability to classify simple problems according to type distinguishes the talented from the average novice (Kahney, 1983) and that the ability to retrieve abstract solution types is an important aspect of learning to program (Anderson *et al.*, 1984). Studying initial problem representation in terms of problem categories is limited because most real-world programming problems are too complex to be assigned to a single category and the study of programming problem categories is more relevant to the later design and planning processes, when smaller subproblems are identified and when design strategies form the basis of categories (Hoc, 1988).

For practical reasons, this aspect of programming (understanding and representing the programming problem) has not been much studied; researchers have even tried to choose programming problems to study for which little specialized domain knowledge is required (e.g. Jeffries *et al.*, 1981; Pennington, 1987a,b). However, there is ample evidence in other problem-solving domains and some evidence in programming research that how the problem is understood and represented is of critical importance in how easily it is solved, and in the correctness of solution (Hayes and Simon, 1977; Larkin, 1983; Kintsch and Greeno, 1985). Secondly, since programming may be seen, in part, as successive transformations of the external problem domain representation into the programming language representation (Brooks, 1983), the relations between domains will critically determine the difficulty of the programming task.

3 Program design and the representation of programming plans

Program design plays a central role in the programming process. Design in this context is often considered to be synonymous with planning, that is, laying out at some level of abstraction, the pieces of the solution and their interrelations. Because plans are structures that are subject to inspection and reorganization, planning serves the dual functions of preventing costly mistakes and simplifying problem solving by providing an overall structure for the problem solution (Anderson, 1983). For programming, most researchers and practitioners agree that this is an important step in determining the quality of the final program, particularly for large or complex programs. In this phase, design strategies and design knowledge are co-ordinated to map the problem representation into a program plan (see Figure 1). Design occurs at both a *general* design level in which requirements and specifications are decomposed into a system structure and at a *detailed* level in which algorithms to implement different modules are selected or created (Yau and Tsai, 1986).

The design of software, as with the design of any complicated artifact, presents challenging problems for the developer. One particular source of difficulty results from the fact that the design subtask interacts with other programming subtasks. That is, as we have already noted, design will alternate with work on problem understanding, coding and revision. One reason this occurs is because of the high degree of uncertainty and incomplete information that is typical of a large-scale programming project. Uncertainty may have many causes but some examples include changing requirements or interfacing technologies under simultaneous development. Information may be incomplete when clients are not clear about their own goals, developers are inventing new products or no one has experience with the application area. A second reason for alternating among subtasks is that decisions made in later subtasks, and problems discovered in later subtasks, may alter decisions made in previous subtasks, necessitating a return to problem understanding or early design phases.

The dominant view of planning discussed in the psychology and artificial intelligence literatures is one of step-wise refinement (Sacerdoti, 1977), in which the primary process is one of top-down, breadth-first decomposition. In this method, a complex problem is decomposed into a collection of (ideally) non-overlapping subproblems. The subproblems are decomposed into further subproblems and this is repeated until the subproblems are simple enough to be solved by retrieving or specifying a known plan for solution (Wirth, 1974). There is some empirical support for software design by step-wise refinement (Jeffries *et al.*, 1981; Carroll *et al.*, 1979; Adelson and Soloway, 1988) and some automatic programming models design in this way (Miller and Goldstein, 1977). However, as a view of what the design subtask actually involves, design by step-wise refinement presents an overly simple view.

First, there is evidence to suggest that the design process is not as orderly as that required by step-wise refinement. Miller and Goldstein (1977) found that in many instances their computer coach needed a mechanism to alter the coach's approved (orderly) expansion. Other data also suggest that there is some amount of alternation between levels of planning as early decisions have implications for later steps and later steps may call into question some aspects of earlier decompositions (Atwood and Jeffries, 1980; Ratcliffe and Siddiqi, 1985). Some are even more pessimistic, suggesting that good programmers 'leap intuitively ahead, from stepping stone to stepping stone, following a vision of the final program; and then they solidify, check, and construct a proper path' (Green 1980, p. 306).

This evidence is more consistent with a second view of planning called opportunistic planning (Hayes-Roth and Hayes-Roth, 1979) in which the plan exists at different levels of abstraction simultaneously and the planner continually alternates between levels. The characterization of planning in this view is multidirectional rather than top-down since observations that arise from planning at lower levels may guide planning at a more abstract level. Evidence from a variety of planning domains suggests that alternation between levels of planning, coping with interdependent subgoals, and opportunism in planning are frequently observed (Rowe, 1987). Opportunism in planning has been explicitly described in work by Visser (1987, 1988; Chapter 3.3) and by Guindon (Guindon *et al.*, 1987). These researchers present ample evidence that software designers skip between levels of detail in design development. It is likely that step-wise refinement methods are used when problems are familiar and of reasonable size, and that more complex and novel design problems must of necessity involve opportunism and less-balanced design.

To describe the design subtask in terms of step-wise refinement also glosses over a major design difficulty: deciding what the units of problem decomposition are or ought to be. This, of course, is the topic of most books on programming technique (see e.g. Wirth, 1974; Dahl *et al.*, 1972; Jackson, 1975, 1983; Yourdon and Constantine, 1979), and is a matter of some dispute (Bergland, 1981). If one subscribes to the original structured programming sequencing constructs, then structural decompositions, even at the abstract level, will focus on the scheduling scheme of subprocedures. Jackson (1975, 1983) argues that the process rather than the procedure is appropriate as a fundamental structural component because of its suitability as a modelling medium of the external problem domain. Yourdon and Constantine (1979) stress dataflow decomposition. There is little empirical research on what kinds of decompositions programmers produce in designing complex programs (see Ratcliffe and Siddiqi, 1985; Guindon *et al.*, 1987, for exceptions) and it is reasonable to expect that the nature of program decompositions will be related to the type of programming task, the programming language, the programmer's training and knowledge. Different decompositions, however, will have implications for later ease of comprehension and maintenance of the program (Bergland, 1981), depending on the extent to which different types of relations between design parts such as the purpose or function of a particular plan unit, the structure of data objects, the sequencing of operations (control flow), and the transformations of data objects (data flow) are explicit or obscure (see Green (1980) and Pennington (1987b) for more discussion of multiple decompositions of program plans).

In their pure forms, problem decomposition and step-wise refinement represent 'analytic' approaches to design (Carroll and Rosson, 1985) and, as descriptions, they miss some important features of design problem solving. For example, prototyping methods of design often involve implementing a kernel solution and then adding to the design by increments (Boehm *et al.*, 1984; Ratcliffe and Siddiqi, 1985; Kant and Newell, 1984). Discovery aspects of design involve trial solutions, keeping features that work and discarding those that don't (Carroll and Rosson, 1985). Mental simulation to estimate design effects and interactions is used extensively (Adelson and Soloway, 1988). It is also the case that design often does not start from scratch, but rather, a prior design is used as a starting point and modified (Visser, 1988; Carroll and Rossen, 1985; Pennington, 1988; Silverman, 1985).

Design is also a very knowledge-intensive subtask, utilizing several different kinds of knowledge (see Figure 1). Brooks (1977) has suggested that an expert programmer may have 50 000 to 100 000 chunks of programming knowledge. The nature of these knowledge chunks would be critical in determining problem decompositions in design. Soloway and his colleagues have described this knowledge base as 'plan knowledge'; a program plan is a stereotypical sequence of program actions that accomplish a certain computational goal (Soloway *et al.*, 1988; Rich, 1981; Chapter 3.1). Other kinds of knowledge implicated in design are: knowledge of design and programming conventions and efficiencies (Soloway *et al.*, 1988), knowledge of design methods and strategies (Jeffries *et al.*, 1981; Guindon *et al.*, 1987; Vessey, 1985: Chapter 3.2), and in some cases, knowledge of a program design language (Ramsey *et al.*, 1983).

4 Coding

Coding a program involves translating the most detailed level of the plan formulation into a programming language. In spite of the importance assigned to the ***design*** of the program, some studies of programming have found that planning occupies a relatively small amount of design time compared to the programming phase. For example Visser (1987) found that her subject spent 1 hour planning compared to 4 weeks coding. This implies that, in practice, design is intimately intertwined with coding in contrast to a common assumption that coding doesn't begin until the design is complete. This is especially true in environments in which prototyping design methods are prevalent. Although, in principle, a prototyping method does not preclude 'requirements before design before coding', in practice the implementation of kernel solutions often precedes full understanding of the problem or full design (see Chapter 3.3). It is also true in environments where high uncertainty causes certain design decisions to be held up or to be made only tentatively and coding works around these problem areas.

The coding process has been described as one of symbolic execution, in which a plan element triggers a series of steps through which a piece of code is generated and then the programmer symbolically executes (i.e. mentally simulates) that piece of code in order to assign an effect to it. This effect is compared with the intended effect of the plan element; any discrepancy causes more code to be generated until the intended effect is achieved. Then the next plan element is retrieved and the process continues (Brooks, 1977). In coding, perhaps more than in any other subtask, the creation of segments of code alternates frequently with evaluation, not only for correctness, but for style, efficiency, consistency of notation, etc. (Gray and Anderson, 1987; Visser, 1987). This alternation between composition and evaluation leads inevitably to the reconsideration of design decisions when the implementation in code fails to achieve the intended effects or does so awkwardly.

To the extent that the plan elements are lost from memory or never completely specified, the evolving code may serve to help reconstruct or alter the guiding plan (Green *et al.*, 1987). In addition, the coding process may be interrupted and changed as a result of processes such as symbolic execution and the evaluation of programming effects. Gray (Gray and Anderson, 1987) calls these episodes 'change episodes' and estimates that they occur as frequently as once every minute, giving the coding process a sporadic and halting nature (Green *et al.*, 1987).

It is well established that when people are learning to write computer programs, they will often find an example piece of code to use as a model for the piece of code they wish to write. However, for experienced programmers, coding is most often described as if the code were generated directly from the plan elements, i.e. from scratch. However, observations of programmers coding real (as opposed to toy) programs reveals that *most* coding does not occur from scratch (Visser, 1987; Pennington, 1988). Rather, the programmer finds a program or parts of other programs to use as coding models, or blatantly borrows old code and modifies it. This means that an important part of programming will be the ability to easily access relevant examples of code that are similar to the program under construction.

Throughout the coding process a representation of the program is built up, storing information about the objects (variables, data structures, etc.), their meanings and their properties. In Brooks' theory and in other speculations about coding behaviour (Shneiderman, 1980; Barstow, 1979), syntactic knowledge is generally considered to be represented independently of semantic programming knowledge. Syntactic knowledge is thought to include a relatively small number (Brooks (1977), estimates 50 to 154) of coding templates detailing internal statement order and syntax, in contrast to the large size estimates of semantic programming knowledge. It is, however, difficult to separate programming language knowledge from the various kinds of knowledge implicated in design phases. Knowledge that is necessary in the coding subtask also includes knowledge of the system on which the program will be implemented and the constraints that that imposes on the code.

The study of coding also cannot be separated clearly from planning in that a definitional line between refinement steps leading up to coding and the actual translation into code may be somewhat arbitrary. Moreover, certain aspects of coding must be taken into account during the design phase, such as language constraints and hardware constraints. Failure to do this has resulted in the creation of whole programs that cannot be executed on the target hardware because they, for example, exceed memory restrictions, or depend on transmission speeds not supported by the device (Krasner *et al.*, 1987).

At one or more points in the development of software, the code must be systematically tested for its correctness. Testing may be performed by the person who coded the program, or may be performed by a separate person or group. It is usually impossible to explicitly test the program on all possible combinations of inputs to the program that might occur in practice, but this is the goal. The development of and knowledge about testing methods and test data generation is a software development skill quite distinct from coding or design (see Chapter 4.2). It is the difficulty of testing for program correctness, among other things, that has led some to advocate formal proofs of correctness as an essential step in program development. Unfortunately, these methods have not yet proved practicable for programs of any realistic complexity.

5 Program maintenance subtasks

Program maintenance subtasks include debugging and modification. Both of these subtasks rely heavily on program comprehension (Jeffries, 1982; Gugerty and Olson, 1986; Nanja and Cook, 1987; Littman *et al.*, 1986). Thus we will also include program comprehension as a maintenance subtask although we believe it is a more fundamental process crossing all subtasks as shown in Figure 1. Green (1980, p. 307) writes, 'Understanding is what it's all about'. In spite of this, we do not have a complete picture of what is involved in program comprehension; as a consequence, our understanding of debugging and modification is similarly limited.

5.1 Program comprehension

Understanding a program involves assigning meaning to a program text, more meaning than is literally 'there'. A programmer must understand not only what each program statement does, but also the execution sequence (control flow), the transformational effects on data objects (data flow), and the purposes of groups of statements (function) (Pennington, 1987a,b). In order to do this, the programmer will employ a comprehension strategy that co-ordinates information 'in the program text' with the programmer's knowledge about programs and the application area. This results in a mental representation of the program meaning. A variety of strategies and representations have been suggested.

It has been suggested that program comprehension involves successive recodings of groups of program statements into increasingly higher-level semantic structures (Shneiderman, 1980; Basili and Mills, 1982). Thus, patterns of operation are recognized as higher-order chunks; patterns of chunks are recognized as algorithms, and so on. This view suggests that a hierarchical internal semantic representation of the program is built during program comprehension from the bottom up. An alternate view is that comprehension is a process of hypothesis testing and successive refinement (Brooks, 1983). In this view, the meanings of the program begin to be built at the outset. For example, even with the name of the program (e.g. main file update) the programmer will have hypotheses about the general function of the program and its major constituents. This implies that the programmer accesses and activates a high-level program schema which partially guides a search for evidence about the expected program components. During the search, the programmer will generate subsidiary hypotheses until they can be matched against beacons (Brooks, 1983) which may confirm parts of the programmer's hypotheses, further refine them or suggest alternatives. Beacons may be procedure names, variable names, or stereotypical code sequences; for example, a particular manner of exchanging values in an array may be recognized as indicating a sort routine and thus serve as a beacon. The programmer's internal representation of the program starts at the top with the program's general function. As higher-level hypotheses are refined and divided, plan elements are added to the representation, followed by the integration of lower-level chunks. The comprehension of a program is complete when the lowest-level plan elements can be bound to actual code sequences in the program. Without a doubt, both top-down and bottom-up processes are involved (Pennington, 1987a,b; Curtis *et al.*, 1984; Letovsky, 1986). For example, matching program plan knowledge to code allows the programmer to make inferences about the goals of the program. This in turn leads to further predictions about program contents, a 'top-down' process. In contrast, mental simulation of program effects is a 'bottom-up' process that enables the programmer to reason about the goal-code relations.

The mental representation of program meaning that is constructed by the programmer is thought to be multilevelled. Drawing on current theories of text comprehension, it has been described in terms of two distinct but cross-referenced representations of a text that are constructed in the course of text comprehension (van Dijk and Kintsch, 1983; Kintsch, 1986). The first representation, the *textbase*, includes a hierarchy of representations consisting of a surface memory of the text, a microstructure of interrelations among text propositions and a macrostructure that organizes the text representation. The second representation, the *situation model* is a mental model of what the text is about referentially. Because the program text is fundamentally about computations, we have called the textbase a 'program model'; similarly, the situation model refers to the application domain, and we have called it a 'domain model'. We have further proposed that the program model will be dominated by procedural relations between program parts and the domain model will be dominated by functional relations (Pennington, 1987a). Exceptional performance in program comprehension is associated with the construction of both mental representations and with mappings between them (Brooks, 1983; Pennington, 1987a).

5.2 Debugging and program modification

The term debugging is sometimes used to refer to the combination of testing and debugging, where testing is a systematic search for program errors. Here, we use debugging to refer to the activities involved in locating and repairing errors in programs once they are known to exist. These errors could have been detected either through testing or through feedback from users of the program. Much of the software in use has associated with it a list of 'outstanding bugs' or known malfunctions. Some of these bugs have consequences that are merely annoying to users of the software but others have serious economic and potentially life-threatening consequences. Thus debugging is a major activity and cost in software development and maintenance.

Debugging is a diagnostic task, similar to other diagnostic tasks such as medical diagnosis and electronic troubleshooting. That is, the program displays some 'symptoms' and the debugger must discover the 'disease' that is causing those symptoms, and 'treat' the disease until the program is 'well' or symptom free. As such, program diagnosis (debugging) usually involves the following activities: (a) understanding the program or system of programs being debugged; (b) generating and evaluating hypotheses concerning the problem; (c) repairing the problem; and (d) testing the system, once repaired (Clancey, 1988; Katz and Anderson, 1988; Vessey, 1985, 1986).

Debugging largely involves understanding what a program *is* doing and what it is *supposed* to be doing (Kessler and Anderson, 1986; Gugerty and Olson, 1986; Nanja and Cook, 1987), and experts in debugging spend much longer understanding the program than do novices (Jeffries, 1982). As a consequence, expert programmers appear to develop a fairly complete mental representation (model) of the program and to understand the possibilities for program errors as causal models of error in this context (Vessey, 1985, 1986, 1989; Jeffries, 1982; Gugerty and Olson, 1986). One way to insure adequate comprehension for debugging is to have written the program yourself. However, most debugging will involve diagnosing programs written and/or designed by others. Thus comprehension skill is central to debugging.

A general debugging strategy is to form a hypothesis about what kind of a bug may create the observed symptoms, to search the code for the location of the bug, modify the program, and run the program to see what effect the changes had. A second key skill then is the ability to generate hypotheses about bugs. Some strategies observed in programmers include: using clues in the output, using tests of internal program states, recalling prior bugs that generated symptoms like the current ones, simulation of program parts, and trial and error (Gugerty and Olson, 1986; Gould, 1975).

A more general model of diagnosis has been put forth by Clancey (1988) in the domain of medical diagnosis that may have a clear application to understanding program debugging. One emphasis of this model is to be specific about the kinds of knowledge that the experienced diagnostician might possess, and how these different kinds of knowledge work together to narrow the diagnostic hypotheses under consideration. One important kind of knowledge is knowledge of diagnostic strategies such as elaborating symptoms when symptoms are known to have many possible causes, and such as thinking at an intermediate level of generality in terms of categories of errors rather than in terms of very specific errors (see also Vessey (1985); and Chapter 3.2 on the importance of strategic knowledge). Strategic knowledge, however, must work on a large network of domain-specific relations between diseases (or categories of diseases) and symptoms (or categories of symptoms). Thus the program debugger should possess debugging strategies and a network of knowledge about which kinds of errors and programmer actions result in which kinds of program bugs.

A second key idea pursued in Clancey's (1988) model of diagnosis and elsewhere in work on explanation-based decision making (Pennington and Hastie, 1988), is that diagnosis for complex problems is often not a result of simple associations between symptoms and diseases. Rather, disease (diagnostic category) is understood as the result of a causal process that in turn resulted in the symptoms (evidence). Under this view, the task of the diagnostician is to build an account of a disease process that can account for the symptoms. Thus, a program error would be understood as something that resulted from programmer or designer activity and diagnosis would centre on trying to understand what actions could have produced an error that produced the symptoms. These aspects of debugging need to be better understood.

The greatest difficulty in debugging appears to be diagnosis. Repair of the program presents many fewer problems (Katz and Anderson, 1988), at least in simple programs. In more complex programs, repairs may involve redesign and may have unanticipated consequences due to interactions between program parts or modules.

Program modification may, of course, take place outside of the context of debugging. For example, users may change their minds about what they want the program to do, resulting in added functions. Once again, it has been proposed that the basis of program modification is program understanding (Fjeldstad and Hamlen, 1983; Littman *et al.*, 1986). Furthermore, the style of comprehension has been shown to be strongly related to success in program modification. For example, Littman *et al.* found that programmers using a systematic comprehension strategy rather than an 'as-needed' strategy were successful at performing a required modification. This result was attributed to the completeness of the mental representation that resulted when a systematic comprehension strategy was used. Obviously, program modification will include, in addition to a base of comprehension, most other programming subtasks as well. Extensive modifications will involve changes to design, extensive coding, retesting, and debugging.

6 Interrelations between programming subtasks

The subtasks of programming are not very different from the subtasks of any design task, or of any problem-solving activity that involves planning: the problem must be understood, the solution sketched out at some level of detail, the solution implemented, corrected and/or modified. Strategies of decomposition, pattern recognition, mental simulation, analogy and causal reasoning will be used. However, as we have stressed, the concept of programming as an unbroken progression through subtasks is not descriptive of programming practices for complicated programming projects.

At the level of individual cognition, it is simply difficult to delineate precisely where one cognitive activity finishes and another begins, such as the distinctions between understanding and design, between design and coding, and so on. This is partly due to our lack of knowledge of the exact cognitive mechanisms underlying these activities but it is also due to their interdependence in reality. The activity of design causes the developer to elaborate his or her understanding. The activity of coding may force one to consider the impossibility or inadequacy of design elements.

There is also inevitable programming subtask interaction at the level of the environments in which software development takes place. Software development is usually distributed over one or more teams (see Chapter 4.1) causing problems in communication and shared understanding. Within a single group, shared conceptualizations of the problem, the design, the code, and so on must be built up. External representations in the form of documents (see Figure 1) serve as central recordings of this. These documents, however, are always incomplete in that they assume certain shared knowledge. They are also notoriously slow to change to reflect current thinking, leaving informal communication as the avenue of building shared views. This problem is exacerbated when different groups are assigned different subtasks. In addition to the problem of building a concensus on the problem being solved, this conceptualization must be passed on to the group attacking the next subtask. For example, environments in which requirements are developed by one group and passed on to another group for high-level design need to pass not only an official document but also their often considerable knowledge of the problem domain. Curtis and Walz (Chapter 4.1) point to the need for team members in these instances whose knowledge 'spans the boundaries' of the two subtasks, to alleviate the information transmission problems.

Finally, software development that is not 'routine', i.e. a slight modification of something that has been done before, is usually characterized by high uncertainty and the need for tentative decisions at every phase of development. This element of the environment is pervasive and of great influence, yet people seem to treat it each time as if it is an aberration and not 'typical' (Krasner *et al.*, 1987). It is a fact of the world that people change their minds about what they want, especially when they see it or use it, and software has already come to be seen as evolving in this sense. However, it is also not uncommon for a software project to require, for one of its components, a piece of software or hardware that is being developed concurrently within the same company or even by another company or government agency. Sometimes features of this import are unknown, or change radically in the course of development, causing reverberating alterations in the development of anything connected to it. Software development and the tasks of programming must be viewed above all within this context of high uncertainty and incomplete knowledge.

References

Adelson, B. and Soloway, E. (1985). The role of domain experience in software design. *IEEE Transactions on Software Engineering,* **11,** 1351-1360.

Adelson, B. and Soloway, E. (1988). A model of software design. *In* M. Chi, R. Glaser and M. Farr (Eds), *The Nature of Expertise.* Hillsdale, NJ: Erlbaum, pp. 185-208.

Anderson, J. R. (1983). *The Architecture of Cognition.* Cambridge, MA: Harvard University Press.

Anderson, J. R., Farrell, R. and Sauers, R. (1984). Learning to program in LISP. *Cognitive Science,* **8,** 87-129.

Atwood, M. E. and Jeffries, R. (1980). *Studies in plan construction I: Analysis of an Extended Protocol,* Technical Report SAI-80-028-DEN, Englewood, Colorado: Science Applications.

Barstow, D. R. (1979). *Knowledge-based Program Construction.* New York: North Holland.

Barstow, D. R. (1985). Domain-specific automatic programming. *IEEE Transactions on Software Engineering,* **11,** 1321-1336.

Basili, V. R. and Mills, H. D. (1982). Understanding and documenting programs. *IEEE Transactions on Software Engineering,* **8,** 270-283.

Bergland, G. D. (1981). A guided tour of program design methodologies. *Computer,* **October,** 18-37.

Boehm, B. W., Gray, T. E. and Seewaldt, T. (1984). Prototyping versus specifying: a multiproject experiment. *IEEE Transactions on Software Engineering,* **10,** 290-302.

Brooks, R. (1977). Towards a theory of the cognitive processes in computer programming. *International Journal of Man-Machine Studies,* **9,** 737-751.

Brooks, R. (1983). Towards a theory of the comprehension of computer programs. *International Journal of Man-Machine Studies,* **18,** 543-554.

Carroll, J. M. and Rosson, M. B. (1985). Usability specifications as a tool in iterative development. *In* H. R. Hartson (Ed.), *Advances in Human-Computer Interaction,* vol. 1. Norwood, NJ: Ablex.

Carroll, J. M., Thomas, J. C. and Malholtra, A. Clinical-experimental analysis of design problem solving. (1979). *Design Studies,* **1,** 84-92.

Carroll, J. M., Thomas, J. C. and Malholtra, A. (1980). Presentation and representation in design problem solving. *British Journal of Psychology,* **71,** 143-153.

Clancey, W. J. (1988). Acquiring, representing, and evaluating a competence model of diagnostic strategy. *In* M. Chi, R. Glaser and M. Farr (Eds), Hillsdale, NJ: Erlbaum.

Cummins, D. D., Kintsch, W., Reusser, K. and Weimer, R. (1988). The role of understanding in solving word problems. *Cognitive Psychology,* **20,** 405-438.

Curtis, Forman, Brooks, Soloway and Ehrlich. (1984). Psychological perspectives for software science. *Information Processing and Management,* **20**, 81-96.

Dahl, O. J., Dijkstra, E. and Hoare, C. A. R. (1972). *Structured Programming.* London: Academic Press.

Fjeldstad, R. K. and Hamlen, W. T. (1983). Application program maintenance study – Report to our respondents. *In* G. Parikh and N. Zvegintzov (Eds), *Tutorial on Software Maintenance.* Silver Spring, Maryland: IEEE Computer Society Press.

Goldman, N., Balzer, R. and Wile, D. (1977). The use of a domain model in understanding informal process descriptions. *Proceedings of the Fifth Joint Conference on Artificial Intelligence.*

Gould, J. D. (1975). Some psychological evidence on how people debug computer programs. *International Journal of Man-Machine Studies,* **7**, 151-182.

Gray, W. D. and Anderson, J. R. (1987). Change-episodes in coding: When and how do programmers change their code? *In* G. M. Olson, S. Sheppard and E. Soloway (Eds), *Empirical Studies of Programmers: Second Workshop.* Norwood, NJ: Ablex.

Green, T. R. G. (1980). Programming as a cognitive activity. *In* H. T. Smith and T. R. G. Green (Eds), *Human Interaction with Computers.* New York: Academic Press.

Green, T. R. G., Bellamy, R. K. E. and Parker, M. (1987) Parsing and gnisrap*: A model of device use. *In* G. M. Olson, S. Sheppard, and E. Soloway (Eds), *Empirical Studies of Programmers: Second Workshop.* Norwood, NJ: Ablex.

Greeno, J. G. and Simon, H. A. (1988). Problem solving and reasoning. *In* R. C. Atkinson, R. J. Herrnstein, G. Lindzey and R. D. Luce (Eds), *Stevens Handbook of Experimental Psychology,* vol. 2, pp. 589-672.

Gugerty, L. and Olson, G. M. (1986). Comprehension differences in debugging by skilled and novice programmers. *In* E. Soloway and S. Iyengar (Eds), *Empirical Studies of Programmers.* Norwood, NJ: Ablex.

Guindon, R., Krasner, H. and Curtis, B. (1987). Breakdowns and processes during the early activities of software design by professionals. *In* G. M. Olson, S. Sheppard and E. Soloway (Eds), *Empirical Studies of Programmers: Second Workshop.* Norwood, NJ: Ablex.

Hayes, J. R. and Flower, L. S. (1980). Identifying the organization of writing processes. *In* L. W. Gregg and E. R. Steinberg (Eds), *Cognitive Processes in Writing.* Hillsdale, NJ: Erlbaum.

Hayes, J. R. and Simon, H. A. (1977). Psychological differences among problem isomorphs. *In* N. J. Castellan, D. B. Pisoni and G. R. Potts (Eds), *Cognitive Theory,* vol. 2. Hillsdale, NJ: Erlbaum.

Hayes-Roth, B. and Hayes-Roth, F. (1979). A cognitive model of planning. *Cognitive Science,* **3**, 275-310.

Hoc, J.-M. (1988). *Cognitive Psychology of Planning.* London: Academic Press.

Jackson, M. A. (1975). *Principles of Program Design.* London: Academic Press.

Jackson, M. A. (1983). *System Development.* New York: Prentice-Hall.

Jeffries, R. (1982). *A comparison of the debugging behavior of expert and novice programmers.* Paper presented at the 1982 meetings of the American Educational Research Association.

Jeffries, R., Turner, A. A., Polson, P. G. and Atwood, M. E. (1981). The processes involved in designing software. *In* J. R. Anderson (Ed.), *Cognitive Skills and their Acquisition.* Hillsdale, NJ: Erlbaum.

Kahney, H. (1983). Problem solving by novice programmers. *In* T. R. G. Green, S. J. Payne, and G. C. van der Veer (Eds), *The Psychology of Computer Use.* London: Academic Press.

Kant, E. and Newell, A. (1984). Problem solving techniques for the design of algorithms. *Information Processing and Management,* **28**, 97-118.

Katz, I. R. and Anderson, J. R. (1988). Debugging: an analysis of bug-location strategies. *Human-Computer Interaction,* **3**, 351-399.

Kessler, C. M. and Anderson, J. R. (1986). A model of novice debugging in LISP. *In* E. Soloway and S. Iyengar (Eds), *Empirical Studies of Programmers.* Norwood, NJ: Ablex.

Kintsch, W. (1986). Learning from text. *Cognition and Instruction,* **3**, 87-108.

Kintsch, W. and Greeno, J. G. (1985). Understanding and solving word arithmetic problems. *Psychological Review,* **92**, 109-129.

Krasner, H., Curtis, B. and Iscoe, N. (1987). Communication breakdowns and boundary spanning activities on large programming projects. *In* G. M. Olson, S. Sheppard and E. Soloway (Eds), *Empirical Studies of Programmers: Second Workshop.* Norwood, NJ: Ablex.

Larkin, J. H. (1983). The role of problem representation in physics. *In* Gentner and A. L. Stevens (Eds), *Mental Models.* Hillsdale, NJ: Erlbaum.

Letovsky, S. (1986). Cognitive processes in program comprehension. *In* E. Soloway and S. Iyengar (Eds), *Empirical Studies of Programmers.* Norwood, NJ: Ablex.

Littman, D. C., Pinto, J., Letovsky, S. and Soloway, E. (1986). Mental models and software maintenance. *In* E. Soloway and S. Iyengar (Eds), *Empirical Studies of Programmers.* Norwood, NJ: Ablex.

Malhotra, A., Thomas, J. C., Carroll, J. M. and Miller, L. A. (1980). Cognitive processes in design. *International Journal of Man-Machine Studies,* **12**, 119-140.

Miller, M. L. and Goldstein, I. P. (1977). Structured planning and debugging. *Proceedings of the Fifth International Joint Conference on Artificial Intelligence.* Cambridge, MA.

Nanja, M. and Cook, C. R. (1987). An analysis of the on-line debugging process. *In* G. M. Olson, S. Sheppard and E. Soloway (Eds), *Empirical Studies of Programmers: Second Workshop.* Norwood, NJ: Ablex.

Pennington, N. (1987a). Comprehension strategies in programming. *In* G. M. Olson, S. Sheppard and E. Soloway (Eds), *Empirical Studies of Programmers: Second Workshop.* Norwood, NJ: Ablex.

Pennington, N. (1987b). Stimulus structures and mental representations in expert comprehension of computer programs. *Cognitive Psychology,* **19**, 295-341.

Pennington, N. (1988). *Design-by-example: a case study.* Unpublished manuscript.

Pennington, N. and Hastie, R. (1988). Explanation-based decision making: effects of memory structure on judgment. *Journal of Experimental Psychology: Learning, Memory, and Cognition,* **14**, 521-533.

Ramsey, H. R., Atwood, M. E. and van Doren, J. R. (1983). Flowcharts versus program design languages: An experimental comparison. *Communications of the ACM,* **26** 445-449.

Ratcliffe, B. and Siddiqi, J. A. (1985). An empirical investigation into problem decomposition strategies used in program design. *International Journal of Man-Machine Studies,* **22**, 77-90.

Rich, C. (1981). *Inspection methods in programming.* Technical Report No. 604, MIT AI Laboratory.

Rowe, P. G. (1987). *Design Thinking.* Cambridge, MA: MIT Press

Sacerdoti, E. D. (1977). *A Structure for Plans and Behavior.* New York: Elsevier.

Shneiderman, B. (1980). *Software Psychology.* Cambridge, MA: Winthrop Publishers.

Silverman, B. G. (1985). The use of analogs in the innovation process: A software engineering protocol analysis. *IEEE Transactions on Systems, Man, and Cybernetics,* **15**, 30-44.

Soloway, E., Adelson, B. and Ehrlich, K. (1988). Knowledge and processes in the comprehension of computer programs. *In* M. Chi, R. Glaser and M. Farr (Eds), *The Nature of Expertise,* Hillsdale, NJ: Erlbaum, pp. 129-152.

Thomas, J. C. and Carroll, J. M. (1979). The psychological study of design. *Design Studies,* **1**, 5-11.

van Dijk, T. A. and Kintsch W. (1983). *Strategies of Discourse Comprehension.* New York: Academic Press.

Vessey, I. (1985). Expertise in debugging computer programs: a process analysis. *International Journal of Man-Machine Studies,* **23**, 459-494.

Vessey, I. (1986). Expertise in debugging computer programs: an analysis of the content of verbal protocols. *IEEE Transactions on Systems, Man, and Cybernetics,* **16**, 621-637.

Vessey, I. (1989). Toward a theory of computer program bugs: an empirical test. *International Journal of Man-Machine Studies,* **30**, 23-46.

Visser, W. (1987). Strategies in programming programmable controllers: a field study on a professional programmer. *In* G. M. Olson, S. Sheppard and E. Soloway (Eds), *Empirical Studies of Programmers: Second Workshop.* Norwood, NJ: Ablex.

Visser, W. (1988). *Giving up a Hierarchical Plan in a Design Activity.* Research Report No. 814, INRIA, France.

Weiser, M. and Shertz, J. (1983). Programming problem representation in novice and expert programmers. *International Journal of Man-Machine Studies,* **19**, 391-398.

Wirth, N. (1974). On the composition of well-structured programs. *Computing Surveys,* **6**, 247-259.

Yau, S. S. and Tsai, J. J. P. (1986). A survey of software design techniques. *IEEE Transactions on Software Engineering,* **12**, 713-721.

Yourdon, E. and Constantine, L. L. (1979). *Structured Design.* Englewood Cliffs, NJ: Prentice-Hall.

Chapter 1.4

Human Cognition and Programming

Tom Ormerod

Department of Human Sciences, Loughborough University of Technology, Loughborough, Leicestershire LE11 3TU, UK

Abstract

This chapter presents an overview of cognition, and reviews the factors that dictate the nature of cognitive models of programming. The computational metaphor is outlined, and the following issues central to information processing are described: knowledge representation, schemas, production rules, procedural and declarative knowledge, attentional and memory resources, semantic memory, problem solving, skill acquisition, and mental models. These issues are fundamental to psychological models of programming presented in later chapters.

1 The relationship between cognition and programming

The fields of cognition and programming are related in three main ways. First, cognitive psychology is based on a 'computational metaphor', in which the human mind is seen as a kind of information processor similar to a computer. Secondly, cognitive psychology offers methods for examining the processes underlying performance in computing tasks. Thirdly, programming is a well-defined task, and there are an increasing number of programmers, which makes it an ideal task in which to study

Psychology of Programming
ISBN 0-12-350772-3

cognitive processes in a real-world domain. This chapter presents an overview of cognition, and reviews the factors which constrain cognitive performance and which dictate the nature of cognitive models of programming.

1.1 The computational metaphor

Cognitive psychology is the study of the mechanisms by which mental processes are carried out, and the kinds of knowledge required for each process. Cognitive processes include perception, attention, memory storage and retrieval, language production and understanding, problem solving and reasoning. These are all operations performed on information by a central processing unit (CPU) with associated memory, to produce an appropriate output. Like a computer, information flows through stages of cognitive processing and storage to give a response output. Also like a computer, information is represented in symbolic form. Recent theories of information processing retain the concept of information in the form of symbols being stored and operated upon before output of a response, although their components are more interactive and less stage dependent than those of earlier theories (e.g. Broadbent, 1958).

Pylyshyn (1984) argues that, since both human and computer output result from operations carried out on symbols, cognition is literally a kind of computation. A computer system can be described at three levels: the software; features of the implementation of programs on the hardware (e.g. the disk-operating system, CPU speed, RAM cache, etc.); and the hardware itself (e.g. circuit boards, keyboard input, etc.). Adopting a literal interpretation of the computational metaphor, similar levels of software, implementation and hardware can be used to describe a cognitive system (e.g. Anderson, 1987a). The 'cognitive software' consists of the mental procedures and knowledge representations used in performing cognitive tasks. For example, strategies such as means-ends-analysis are equivalent to programs for undertaking certain problem-solving tasks. The 'cognitive implementation' of software concerns the mechanisms for carrying out mental procedures and knowledge representation, such as symbol manipulation, storage, retrieval and so on. For example, limitations of attention and memory constrain the knowledge and procedures available during problem solving. Anderson (1987a) argues that studying cognitive implementation, despite identifying constraints on cognitive performance, is both more difficult and less rewarding than studying cognitive software. Although this is contentious, it is consistent with research into programming which is mainly concerned with identifying the mental procedures and knowledge used in programming.

The 'cognitive hardware' consists of the physiological structures, notably the brain, upon which cognitive processes are implemented. Physiological structures are not a central concern of cognitive psychology, since one of the major assumptions underlying the computational metaphor is that cognitive processes are determined by their function, and not by the structure of the hardware upon which they are based. The computational metaphor dominates cognitive psychology, although some argue that accounting for cognition in terms of symbol manipulation is implausible, and that the biological basis of cognition should not be ignored (e.g. Searle, 1980). A recent development, which offers an alternative to the computational metaphor for studying cognition, is 'connectionism' (e.g. Rumelhart and McClelland, 1986). Connectionist modelling is closer to the level of cognitive hardware, in that it uses the parallel processing capabilities of the brain to explain how cognitive processes are carried out. Connectionist models are not of direct relevance to programming research because they focus on cognitive hardware rather than software. However, future developments in parallel computer architectures may include concepts such as constraint satisfaction programming (see Chapter 1.2). The cognitive skills necessary to cope with programming a parallel-processing machine may therefore be an important topic for future research.

1.2 Important themes in cognition

A division is often made between knowledge representation and the processes which operate on representations of knowledge. At the level of cognitive implementation, theories of representation describe the form in which individual units of knowledge are stored. Pylyshyn (1979) claims that all information is represented as propositions, which are language-like assertions retaining only the meaning of individual items of knowledge, rather than representing knowledge in the exact form in which it arrives as perceptual input. Alternatively, Paivio's (1971) 'dual code' theory suggests that verbal information is represented in linguistic form, and visual information is represented in spatial image or analogue form. Analogue representations may determine performance in some programming tasks. For example, Ormerod *et al.* (1986) found that different reasoning strategies were used in comprehending diagrammatic and list representations of Prolog clauses. More important to programming research, however, is the organization of knowledge at the level of cognitive software (for example, the schema and strategy accounts of programming knowledge discussed in Chapters 3.1 and 3.2).

1.2.1 Schema representations

An important approach to describing knowledge representation at a cognitive software level is the 'schema'. A schema (after Bartlett, 1932) consists of a set of propositions that are organized by their semantic content. A perceptual input, cognitive goal or output of a cognitive process may evoke a schema with a related semantic content. Once evoked, it provides an organized body of knowledge which is appropriate for the task in hand. For example, Figure 1 shows a schematic representation of part of the knowledge a programmer may possess about programming languages. If a programmer is required to think of an alternative to iteration for dealing with lists, the words 'iteration' and 'lists' will activate the schema 'programming languages'. This will allow the programmer to access knowledge about looping constructs, and thus generate alternatives such as 'recursion'. There are two basic principles of schema theories: first, that cognitive processing is guided and limited by the application of prior knowledge; and secondly, that schemas contain relatively abstract knowledge which is independent of any one event. Thus, a schema highlights relevant information in a task domain, and may add new information when there is insufficient information in a task domain. Figure 1 illustrates a number of components of schemas, such as hierarchical organization, default values for generating information which is not present in the task domain, and slots which can be filled by matching information from the task domain. A schema offers a method for limiting the amount of inputed information, or 'bottom-up' control, that is needed to perform a task. Schemas provide 'top-down' control, using prior knowledge to restrict the operations that may be undertaken.

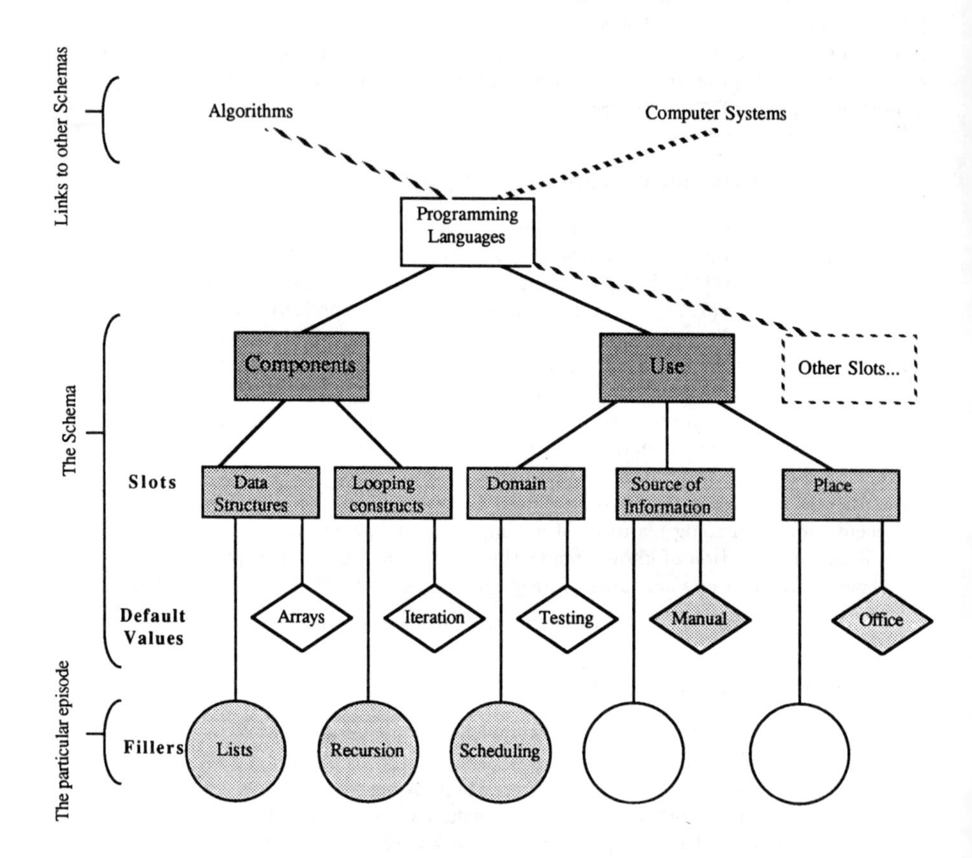

Figure 1: The schema 'programming languages' represents part of the general knowledge a person may have about programming languages. This schema is linked to others (e.g. 'algorithms'). The particular episode represents an encounter with a previously unknown language (such as Lisp). The shaded areas represent information available for a particular task. Information may originate from the task domain as 'fillers' for the schema slots, or from the schema as 'default values' where no conflicting task domain information is present. This representation of schemas, based on one devised by Cohen *et al.* (1986, p. 28), is only one of many possible representations. For example, the 'script' shown in Figure 3. is also a schematic representation.

Schema theories appear in various guises, such as 'frames' (Minsky, 1975), which are schemas originally employed in artificial intelligence (AI) models of vision. Schank and Abelson (1977) identify a particular kind of schema known as the 'script'. A script is a sequence of abstracted actions which occur in common events, with slots for specific instances. For example, a 'restaurant' script contains a sequence of general scenes such as entering, seating, ordering, paying and leaving, along with specific exceptions (e.g. the time someone shouted for service). Schank and Abelson (1977) also identify 'plans', which determine the inferences required for understanding situations for which there are no stereotypical event sequences. Scripts describe action sequences which have occurred many times (e.g. visiting a restaurant), whereas plans describe the production of novel action sequences (e.g. robbing a liquor store). The inferences determined by a script are based on prior knowledge, whereas plan inferences are based on achieving goals in the absence of task-specific knowledge. The processes involved in 'planning' have been described in a number of AI models (e.g. Wilensky, 1981), though there have been surprisingly few empirical investigations of planning. Somewhat confusingly, the programming plans (Soloway and Erlich, 1984) discussed in Chapter 3.2 are closer to scripts of programming knowledge.

Schema theories have been used to account for a large number of cognitive processes, such as memory storage and retrieval (e.g. Rumelhart and Norman, 1983), language understanding (e.g. Schank and Abelson, 1977) and problem solving (e.g. Gick and Holyoak, 1983). There is a danger of the term 'schema' being too nebulous to be of use, and there is a tendency to use the terms 'schema' and 'knowledge' interchangeably. Therefore, it is important for schema theories of programming knowledge to specify explicitly the mechanisms which mediate the acquisition and use of knowledge. Also, a schema theory must be descriptive rather than prescriptive, In other words, it must model the knowledge that programmers really use rather than the knowledge they ought to use.

1.2.2 Production rule representations

Another representation is the 'production rule'. Production rules emphasize the process aspects of cognition, whereas schemas emphasize the representational aspects. A production rule consists of two propositions forming a 'condition-action' pair, one of which is a goal or desired cognitive state, the other being the action or subgoals required to achieve that state. Figure 2 shows examples of some production rules used in Lisp programming (Anderson *et al.*, 1984). If the conditions of a production rule are satisfied by matching them with a perceptual input, retrieval of knowledge from a long-term store, or by the action of another rule succeeding, then the rule is fired or 'activated' (i.e. made accessible to conscious attention) and its action is undertaken. The successful firing of a production rule represents a cognitive step. As an example from Figure 2, a novice learning Lisp by solving textbook exercises might use the production rule P1 as the first step in writing a new function. The conditions would be matched by the existence of a goal to write a function and the prescence of a previous example in the textbook, and the actions to compare the example to the required function and then map the example's solution onto the present problem would be carried out. Production rules are described by Anderson (1987a) as a programming language for cognitive software, in that sets of rules combine to give a 'production system' for undertaking a cognitive task, where the action of one rule provides the conditions for firing the next rule. Production systems have been

A general production rule used by novices in the declarative stage

P1: IF the goal is to write a function and there is a previous example
THEN set as subgoals
1. to compare the example to the function
2. map the example's solution onto the current problem.

....

Compilation of a production rule for list processing

Proceduralization

P2: IF the goal is to code a relation defined in an argument
and there is a Lisp function that codes this relation
THEN use function with the argument and set as a subgoal to code the argument.

Becomes;

P3: IF the goal is to code the first member of a list
THEN use CAR of the list and set as a subgoal to code the list.

Composition

Given another production rule P4;

P4: IF the goal is to add an element to a list
THEN use CONS on the element and the list
and set as subgoals to code the element and to code the list

Adding P3 to P4 gives;

P5: IF the goal is to add the first member of one list to another list
THEN CONS the CAR of the first list to the second list
and set as subgoals to code the first list and to code the second list

....

A specific production rule for experts in the procedural stage

P6: IF the goal is to check that a recursive call to a function will terminate
and the recursive call is in the context of a MAP function
THEN set as a subgoal to establish that the list provided to the MAP function
will always become NIL after some number of recursive calls.

Figure 2: Some possible production rules underlying Lisp programming skills at different stages of skill acquisition, based on Anderson *et al.* (1984). In each production rules the 'IF ...' antecedents give the conditions for the rule to be fired, and the 'THEN ..' consequents give the actions that occur if the consequents are met.

proposed to underlie many cognitive processes, notably those involved in problem solving (e.g. Anzai and Simon, 1979).

Production rules are frequently used to represent 'procedural' knowledge, that is knowledge about *how* to carry out a cognitive task, whereas schemas are often used to represent 'declarative' knowledge, that is knowledge about *what* constitutes a task domain. This parallels a distinction (described in Chapter 1.1) between declarative languages (e.g. Prolog) and procedural languages (e.g. Pascal). It has been claimed that a declarative style is more natural than a procedural style of knowledge representation, and hence that declarative programming languages have advantages over procedural languages (e.g. Miller, 1974). However, others suggest that the preferred style is context dependent (e.g. Rumelhart and Norman, 1983; Gallotti and Ganong, 1985). A related issue is the extent to which a person has conscious access to their knowledge. Whilst declarative knowledge is available to conscious inspection and control, procedural knowledge is applied automatically without conscious control. Again, the distinction is not absolute, and depends on the task and level of expertise involved. The issue of control over cognitive processing has a number of implications for programming, and is discussed in the next section.

2 Constraints on cognitive skills

Programming is a 'high-level' cognitive task, in that it involves problem solving and linguistic skills which require a number of 'lower-level' cognitive tasks to be carried out at the same time. These include the perception, attention to relevant aspects, and short-term storage of task information, and the retrieval of relevant long-term knowledge. Performance in these tasks dictates a number of constraints on programming performance.

2.1 Resources for cognitive processing

Increasingly, theories viewing attention and short-term memory as isolated operations which restrict perceptual input (e.g. Broadbent, 1958) have been replaced by theories of limited 'resources' which restrict both input and output. For example, early theories of memory (e.g. Atkinson and Shiffrin, 1968) identified a passive short-term store with a limited storage capacity of approximately seven items (Miller, 1956), and a long-term store of unlimited capacity. Recent accounts maintain the distinction between short- and long-term stores, but suggest a more active role for short-term memory. Baddeley and Hitch (1974) propose a 'working memory' model of short-term memory. The main component is a central executive, which allots limited resources to either storage or processing. The central executive is served by sub-systems for rehearsal of verbal and spatial information. A working memory system is embodied in a number of cognitive theories (e.g. Anderson, 1983).

The exact nature of cognitive resources is unspecified, but the term implies a limited availability of conscious effort and storage capacity. The performance of two demanding tasks at the same time is difficult, not because information from only one task is available, but because the cognitive system lacks the resources to perform both tasks. Since the number of processes involved in programming is large, attentional and memory resources must be divided up amongst competing processes, thereby limiting performance. Resource theories may explain the source of some programming errors. For example, insufficient short-term memory resources have been suggested to account for errors made by novice Lisp programmers (Anderson and Jeffries, 1985). A resource account of programming errors implies that performance can be improved if the resource requirements of a task are decreased. For example, using appropriate perceptual cues in notations may reduce the attentional demands of program comprehension (this is discussed further in Chapter 2.2).

2.1.1 Controlled and automatic processes in attention

Shiffrin and Schneider (1977) propose a theory of 'controlled and automatic processes' in attention. Automatic processes are analogous to compiled programs in that they are carried out directly in terms of the cognitive implementation or hardware, whereas controlled processes are like interpreted program code that requires translation into machine-specific terms at run-time. Controlled processes require attentional resources, are of limited capacity, and can be used in different contexts. Automatic processes do not require attention, are not capacity limited, but like compiled programs are not modifiable. For example, an expert Lisp programmer may write the syntax of a function definition as an automatic process, but may have to use controlled processes to specify the function itself. In the same way that compiling code speeds up the running of a program, repeated practice reduces the attentional demands of a task by allowing previously controlled processes to be automated.

The activation of automatic processes precludes conscious access to cognitive processing, thereby preventing the monitoring of task performance. As a consequence, the inappropriate activation of automatic processes may lead to errors in performance (Reason, 1979). Automated knowledge of programming skills releases resources which can be used for other programming tasks. However, automatic processes are restricted in their application, and their use in programming tasks which differ from the learning domain may lead to errors. For example, a common error made by Prolog novices who are experienced in Lisp programming is to add unnecessary brackets at the beginning of Prolog program clauses. This kind of error is minor if there are sufficient resources for the controlled processes required for debugging, but it becomes more significant when resources are tied up by other processing demands.

2.2 The contents of memory

A number of factors affect the encoding, storage, and retrieval of long-term memories (for a review, see Eysenck, 1984). Encoding is affected by the depth of processing taking place as information is presented, as well as the distinctiveness of the information to be encoded. Storage is affected by the nature of information rehearsal, elaborative rehearsal being more effective than maintenance rehearsal. Retrieval of information has been shown to be a function of the cues available both at encoding and retrieval time, and the nature of the task (e.g. whether recognition or recall is required). A distinction is often made between semantic memories which contain general knowledge about the world, and episodic memories which contain personally encountered events and experiences linked to specific times and places. A strong division between the two memory systems (e.g. Tulving, 1985) is contentious. In programming research, usually only semantic memory is examined, although structures in semantic memory must be constructed from individual episodes. This is important in learning programming, where schematic knowledge of programming concepts is developed by generalizing across a number of repeated learning episodes.

2.2.1 Semantic networks

One of the most important representational systems proposed for long-term memory is the 'semantic network' (Collins and Quillian, 1970). In a semantic network, semantically related propositions are represented as nodes linked together in a hierarchical fashion. So, in a network of semantic knowledge about animals, 'animal' would be above 'bird', which would be above 'robin', and so on. To store information with minimal redundancy, features common to a number of concepts are associated with the highest level they have in common, and lower-level nodes inherit all the features of higher-level nodes. For example, information such as 'can fly' and 'has feathers' would be associated with the 'bird' node, but could be accessed from lower nodes such as 'robin'. A parallel can be drawn between the concept of inheritance in semantic networks and in object-oriented programming languages such as Smalltalk (Goldberg and Robson, 1986), where changes made in parent classes of objects are propagated to subclasses. Inheritance and access to semantically related nodes in semantic networks occur through 'spreading activation' (Collins and Loftus, 1975), where activation of a source node spreads over time to allow retrieval of information embodied in other related nodes (for a discussion see Anderson, 1984). Semantic networks represent a long-term store of declarative knowledge in Anderson's (1983) ACT* model.

2.2.2 Schemas in semantic memory

At a higher level of organization, it has been proposed that semantic memory is organized into schemas (as discussed earlier). Alba and Hasher (1983) identify four features of schemas which dictate the contents of semantic memory. These are selection, abstraction, interpretation and integration. Activation of an existing schema selects new information for encoding, restricting it to salient or atypical events. For example, giving advance information to activate a relevant schema improves the recall of a complex abstract passage (Bransford and Johnson, 1972). Information in schemas is abstracted so that only the semantic content is stored and episodic details are lost, often leading to distortions in retrieval (Bransford *et al.*, 1972). Schemas use default values to make inferences and simplifications, and new information is integrated with the products of these interpretations to update or form new schemas. Alba and Hasher (1983) suggest that evidence for schema models of memory is equivocal. They argue that schemas cannot account for the richness of recall, and that semantic memory must contain an episodic component. However, schemas explain the flexibility of cognitive processing in the absence of complete information. For example, schematic organization accounts for the chunking of expert knowledge found in program recall studies (e.g. Adelson, 1981).

3 Programming as a problem-solving skill

Newell and Simon (1972) identify four features of a problem: the initial state; the goal state; the operators available for moving from the initial state to the goal; and restrictions on the operators. Programming may be seen as a problem-solving activity, where the initial state is the problem for which a program is required, and the goal state is both the solution which the program can calculate, and also the program itself. The operators consist of the syntactic and semantic features of the language and the cognitive skills of the programmer, and the operator restrictions are imposed by limitations of the language and the problem description. Problems are often described as having a 'state space', which consists of all the features of a problem, and all the possible moves that may be made between initial and goal states. This may be contrasted with a 'problem space' (Simon, 1978), which is the mental representation of a problem, together with the problem solver's prior knowledge. A goal of programming research is to examine differences between the state space and the problem space of the programming task.

3.1 Acquisition of cognitive skills

3.1.1 Expert-novice differences

One approach to studying cognitive skills is to compare the performance of novices and experts. Much of the research on expert-novice differences has examined programming, although other domains have been studied (e.g. physics, chess and medicine). Novices and experts differ in many aspects of problem solving, notably in the ways they represent problems, the strategies they choose for tackling problems, and the ways in which their knowledge about problem domains is organized. For example, differences in problem representation have been identified by Weiser and Shertz (1983). They found that novice programmers group programs according to superficial characteristics such as the application area, whereas experts group problems according to algorithms. Larkin (1981) found strategy differences, where novices solving physics problems tend to work backwards from the problem goal, whereas experts work forwards from the problem givens to the goal. Anderson (1985) suggests that in programming novices work forwards, writing a program line by line, whereas experts work backwards, breaking the program goal into modular units. Physics problems have a large number of givens, which are more predictive of the solution than are the problem goals. In programming, the problem givens are programming languages themselves, which are less predictive of the solution than the programming goal.

An expert-novice difference with important implications for programming research is the apparent schematic organization of expert knowledge (discussed further in Chapter 3.1). For example, studies of the recall of chess positions by novices and experts (Chase and Simon, 1973) showed that grandmasters recall game positions from long-term memory as 'chunks' of information. Grandmasters had greater recall of positions from realistic games than novices, but were no better than novices at recalling random positions. Similarly, McKeithen *et al.* (1981) found that experts recalled more lines from real Algol programs than novices, but there was no difference between experts and novices in the recall of scrambled programs. The difference between experts and novices increased over a number of trials, which contrasts with the findings of Chase and Simon (1973), who found that the advantage for experts declined. Grandmasters have chunks representing whole chess boards, but programs are too complex to be represented by individual chunks. Therefore it is necessary to reconstruct a program over a number of trials, by accessing schematic knowledge of programming concepts.

3.1.2 The ACT* model of skill acquisition

Another method of investigating cognitive skills is to study the changes an individual goes through in acquiring a skill. Anderson's (1983) ACT* model of skill acquisition is based on studies of individuals acquiring skills such as programming. Anderson proposes three stages in acquiring a cognitive skill. The first is a 'declarative' stage, where novices apply very general production rules to the declarative information given in a task or stored in long-term memory in order to solve a problem. The declarative stage does not commit the learner to task-specific procedures at too early a stage. However, the demands placed on the capacity of working memory by using general production rules requiring large amounts of declarative knowledge mean that performance is slow and error prone. A second stage of 'knowledge compilation' reduces the demands on the capacity of working memory by developing task-specific rules. This is achieved by 'composition', where production rules are combined into a single rule, and 'proceduralization', where task-specific information is added to production rules. The final stage is 'procedural' learning, where production rules are strengthened by a process of 'tuning', in which the speed and accuracy of rule application increases with practice.

Anderson *et al.* (1984) investigated the acquisition of Lisp programming skills by studying protocols (subjects' verbal reports and keystrokes made during problem solving) of novices learning to write functions. Figure 2 illustrates some examples of production rules which underlie different stages in the acquisition of Lisp programming skills. P1 is a rule applied by novices in problem solving by a process of analogy. A function is constructed by mapping the declarative knowledge of an example solution onto the declarative task information. Anderson (1987a) describes general production rules as 'weak method problem solutions'. These are used when a novice has no task-specific production rules, and must use very general methods to solve novel problems. Another example of a weak method problem solution is working backwards from a given solution. Knowledge compilation is illustrated by rules P2 to P5. Proceduralization of P2 produces a rule P3 which is specialized for operating on the head of a list. The composition of rules P3 and P4 produces a single rule P5 for adding the first element of one list to another list. P6 is an example of a 'tuned' rule that an expert would possess. Knowledge compilation may be characterized as 'success-driven' learning, that is, learning that results from the successful application of production rules. Anderson (1987b) argues that compilation often leads to one-trial learning, and that verbalization of performance is not possible once a procedural stage has been reached. Also, different production rules underlie different procedural tasks carried out in the same declarative domain. For example, McKendree and Anderson (1987) found that the skills acquired in learning to evaluate simple Lisp functions did not transfer to the task of writing Lisp functions.

3.1.3 The 'dynamic memory' approach to skill acquisition

The 'success-driven' learning in ACT* contrasts with 'failure-driven' learning in Schank's (1982) 'dynamic memory' theory, which is an extension of Schank and Abelson's (1977) scripts theory. Figure 3 shows a script of a possible program debugging session. Each event in the script is an instantiation of a 'scene', which is a generally defined sequence of actions. Scenes which commonly occur together are grouped into scripts by 'memory-organization packets' (MOPs) which represent high-

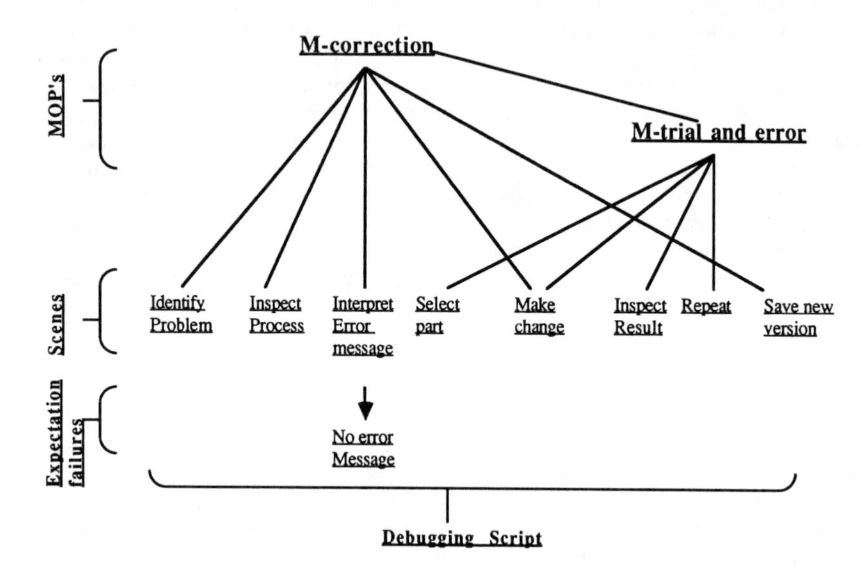

Figure 3: A hypothetical model of knowledge a novice might employ in a program debugging task, based on Schank's 'dynamic memory' theory (Schank 1982, p. 89). The MOPs (Memory Organization Packets) show two different sources of high-level knowledge that might generate expectancies in a debugging task. These MOPs may also guide performance in other tasks. For example, the MOP 'M-correction' might be accessed when mending a car, and the MOP 'M-trial and error' might be accessed when solving a maze puzzle. The five scenes generated by the MOP 'M-correction' represent events that generally occur together when correcting objects that function incorrectly. This MOP is linked to other MOPs which dictate strategies for making changes, such as the MOP 'M-trial and error'. The four scenes generated by the MOP 'M-trial and error' represent a possible novice debugging strategy, which is to generate potential corrections randomly and test them for success. The expectations raised by each scene in the script guide the debugging process. When an expectation fails, the MOP that organizes that scene must be adjusted to account for the failure. For example, the absence of an error message in running a faulty program may indicate a semantic rather than syntactic error. The novice must learn to discriminate between these sources of bugs, by adding additional scenes to the relevant MOPs.

level knowledge about scripted events. Figure 3 illustrates how MOPs which have different semantic contents (e.g. 'correction' and 'trial and error' MOPs) are used in the same task. MOPs control the learning of new skills by generating expectations about events. Learning is 'failure driven' in that it takes place when an expectation fails, under which circumstances the causal chain of events that led up to the failure is traced, and the MOP is then altered. ACT* and dynamic memory theory are not necessarily mutually exclusive, but may describe the learning of different skills. For example, learning to construct program code may be success driven, whereas debugging and testing may require failure-driven learning.

3.1.4 Transfer of problem-solving skills

Limitations of knowledge organization, representation and strategy provide major constraints on the problem space of a problem solver (for a review see Kahney, 1986). The acquisition of expertise overcomes these constraints and allows experts to tackle a broader range of related problems. One mechanism for doing this is to transfer skills from one problem domain to another. Transfer is an increasingly important area of research in problem solving, as well as in programming (discussed in Chapter 2.4). For example, Gick and Holyoak (1980) found that subjects were able to use an analogous example problem with a worked solution to solve Duncker's (1945) 'radiation problem'. Gick and Holyoak (1983) suggest that subjects learn to solve similar problems by using analogical transfer mechanisms, and develop schemas for particular classes of problem by 'schema induction'. Schemas are induced by forming an analogy between two or more training problems and solutions. A schematic representation of an abstracted solution is formed, which can then be evoked by the presence of appropriate context in a new problem.

Not all experiments have found successful transfer. Reed *et al.* (1985) found that transfer occurred only between equivalent algebra problems, and not between similar problems requiring a small manipulation of the problem solution. Positive transfer was found between similar problems, however, when a complex practice problem preceded a more simple transfer problem. This suggests that the transfer attributable to an analogy between equivalent practice and transfer problems is based on a shallow understanding of the problem features. When a deeper understanding of the problem is required, it is necessary to give practice problems where all the steps of the transfer problem solution are explicated. The conditions under which transfer may occur have important implications for training programming skills. For example, negative transfer of performance arising from inappropriate analogizing was found between Prolog comprehension tasks with different thematic content and representation (Ormerod *et al.*, 1990). The real-world familiarity of the training tasks prevented subjects from developing domain-independent comprehension strategies. Similarly, White (1988) found evidence of inappropriate transfer of knowledge by Pascal experts in solving Prolog debugging tasks. In other words, the Prolog task was carried out in terms of the subjects' Pascal knowledge.

3.2 Deductive reasoning

The process of constructing and testing a computer program may be compared to constructing and testing hypotheses by deductive reasoning. Deductive reasoning has traditionally been separated from other problem-solving activities, partly because formal 'competence' models exist for this skill. Piaget amongst others proposed that formal logic underlies human reasoning abilities (for a review see Gross, 1985), such that reasoning is carried by applying the rules of formal logic to propositions, independent of their real-world content or representation, to derive inferences. Thus, a programmer should be able to derive a program specification by formal logical reasoning, a process which should be unaffected by the programming domain or language notation. Indeed, some proponents of logic programming (e.g. Kowalski, 1982) have suggested that languages such as Prolog are closer to human-reasoning processes than conventional languages such as Pascal, because of an assumed match between the formal logic underlying Prolog and human reasoning.

A large body of empirical evidence exists showing systematic biases in reasoning, determined by the thematic content and representation of reasoning tasks (for a review see Evans, 1986). In the context of programming, this evidence is consistent with evidence of the influence of prior experience and problem representation on programming (e.g. Ormerod *et al.*, 1986; White, 1988). It also falsifies the arguments of Kowalski (1982) about the psychological advantages of logic programming languages. Biases in reasoning cannot be accounted for by a theory based on formal logic, but instead require a theorical explanation based on observed performance rather than logical competence. For example, Cheng and Holyoak (1985) found that a 'permission' rule such as 'if you are to drink alcohol, then you must be over eighteen' facilitated logical performance on Wason's (1966) selection task, where subjects must test the truth of the rule by selecting falsifying instances. They suggest that reasoning is carried out, not by logical inferences, but by 'pragmatic reasoning schemas' elicited by statements involving permission, obligation, causation and so on. These are generalizable knowledge structures containing rules for generating useful inferences.

3.2.1 'Mental models' theory

An influential theory which accounts for reasoning and language understanding without recourse to logic is Johnson-Laird's (1983) 'mental models'. In its most general form, a mental model is a mental representation of a problem space. It differs from state space and schematic representations in that, although individual units of information may be represented as propositions, a mental model is constructed out of propositions to form an analogue of the real world representation of a problem. In other words, a mental model has the same functional nature and structure as the system it models. For example, Mani and Johnson-Laird (1982) found that subjects recall the semantic content of spatial descriptions consistent with only one layout (e.g. 'the spoon is to the left of the knife; the plate is to the right of the knife', etc.), but recall verbatim details of spatial descriptions where a number of layouts are possible. This suggests that subjects construct a mental model of determinate descriptions, but resort to rote learning of unconnected propositions in indeterminate descriptions.

Johnson-Laird (1983) offers a 'procedural semantics' for the construction and searching of mental models in working memory, which allows errors in performance to be predicted on the basis of the order in which propositions are added to a mental model, and the number of alternative models which must be constructed. A general use of the term 'mental model', where it has the same relation structure as the real-world domain but a procedural semantics is not specified, is adopted by many theorists in problem solving (e.g. Gentner and Stevens, 1983), and in human-computer interaction (e.g. Manktelow and Jones, 1987). The creation and testing of mental models may explain aspects of expert programming skills, such as the interweaving of a number of programming plans into a single program (e.g. Rist, 1986).

3.3 Psycholinguistics and programming research

The review has for reasons of space been highly selective. The most obvious omission is a discussion of psycholinguistics (for a recent review, see Garnham, 1985). In part this is because many important concepts are covered in other topics, which illustrates how cognitive processes cannot be neatly partitioned as they were in early cognitive models. For example, schematic representations of knowledge have been used to account for text comprehension (Kintsch and van Dijk, 1978). Similarly, mental models theory has been used to account for the comprehension of sentences which require implicit inferences (Garnham, 1987). A competence versus performance debate has also occurred with language as well as reasoning (e.g. Chomsky, 1965). Theories of language understanding tend to emphasize either syntactic analyses (e.g. Fodor *et al.*, 1974) or semantic analyses (e.g. Schank, 1972). Programming research would seem amenable to a similar division of research areas. However, programming tends to be studied as a problem-solving rather than a linguistic activity, with notable exceptions (e.g. Sime *et al.*, 1977). Thus the focus of research into programming is not on syntactic features of different languages, but on semantic programming knowledge. However, as in recent theories of language understanding (e.g. Johnson-Laird, 1983), the interaction between semantic and syntactic components of the programming task has recently been highlighted (e.g. Arblaster, 1982).

4 Conclusions

The review of cognitive psychology presented in this chapter highlights the approaches to understanding human cognition which are of special relevance to programming research. Concepts that recur in many cognitive theories include schemas, production systems, limited resources, automation of skills with practice, working memory, semantic networks and mental models. Most employ propositional representations of one form or another, in which information is represented at a symbolic level.

A number of cognitive theorists (e.g. Fodor, 1983; Marr, 1982) argue that psychologists should concentrate their efforts on understanding processes such as perception, which appear to be carried out effortlessly and without error. They suggest that one should study the processes people are good at before tackling areas such as problem solving where people are slow and error prone. From an applied psychological perspective, it is just these areas where psychological research is at its most useful. Therefore, the motivations for psychological research into programming are strong, both for making programming an easier task, and for adding to a relatively unstable area of psychological theory.

References

Adelson, B. (1981). Problem solving and the development of abstract categories in programming languages. *Memory and Cognition,* **9(4)**, 422-433.

Alba, J.W., and Hasher, L. (1983). Is memory schematic? *Psychological Bulletin,* **93(2)**, 203-231.

Anderson, J. R. (1983). *The Architecture of Cognition.* Cambridge, MA: Harvard University Press.

Anderson, J.R. (1984). Spreading activation. *In* J. R. Anderson and S. M. Kosslyn (Eds.), *Tutorials in Learning and Memory.* New York: Freeman.

Anderson, J. R. (1985). *Cognitive Psychology and its Implications,* 2nd edn. New York: Freeman.

Anderson, J. R. (1987a). Methodologies for studying human knowledge. *Behavioural and Brain Sciences,* **10(3)**, 467-505.

Anderson, J. R. (1987b). Skill acquisition: compilation of weak-method problem solutions. *Psychological Review,* **94(2)**, 192-210.

Anderson, J. R. and Jeffries, R. (1985). Novice LISP errors: undetected losses of information from working memory. *Human-Computer Interaction,* **1**, 107-131.

Anderson, J.R., Farrell, R., and Sauers, R. (1984). Learning to program in Lisp. *Cognitive Science,* **8**, 87-129.

Anzai, Y. and Simon, H. A. (1979). The theory of learning by doing. *Psychological Review,* **86(2)**, 124-140.

Arblaster, A.T. (1982). Human factors in the design and use of computing languages. *International Journal of Man-Machine Studies,* **17**, 211-224.

Atkinson, R. C. and Shiffrin, R. M. (1968). Human memory: a proposed system and its control processes. *The Psychology of Learning and Motivation,* vol 2. New York: Academic Press.

Baddeley, A. and Hitch, H. (1974). Working memory. *In* G.H. Bower (Ed.), *The Psychology of Learning and Motivation,* vol 8. New York: Academic Press.

Bartlett, F. C. (1932). *Remembering: A Study in Experimental and Social Psychology.* London: Cambridge University Press.

Bransford, J. D. and Johnson, M. K. (1972). Contextual prerequisites for understanding: some investigations of comprehension and recall. *Journal of Verbal Learning and Verbal Behavior,* **11**, 717-726.

Bransford, J. D., Barclay, J. R. and Franks, J. J. (1972). Sentence memory: a constructive versus interpretive approach. *Cognitive Psychology,* **3**, 193-209.

Broadbent, D. E. (1958). *Perception and Communication.* Oxford: Pergamon Press.

Chase, W. G. and Simon, H. A. (1973). The minds eye in chess. *In* W. G. Chase (Ed.), *Visual Information Processing.* New York: Academic Press.

Cheng, P. W. and Holyoak, K. J. (1985). Pragmatic reasoning schemas. *Cognitive Psychology,* **17**, 391-416.

Chomsky, N. (1965). *Aspects of the Theory of Syntax.* Cambridge, MA: MIT Press.

Cohen, G., Eysenk, M. W. and Le Voi, M. E. (1986). *Memory: A Cognitive Approach.* Milton Keynes: Open University Press.

Collins, A. M. and Loftus, E. F. (1975). A spreading-activation theory of semantic processing. *Psychological Review,* **82**, 407-428.

Collins, A. M. and Quillian, M. R. (1970). Does category size affect categorization time? *Journal of Verbal Learning and Verbal Behavior,* **9**, 432-438.

Duncker, K. (1945). On problem solving. *Psychological Monographs,* **58(270)**, 1-113.

Evans, J. St. B. (1986). Reasoning. *In* H. Beloff and A. M. Colman (Eds), *Psychology Survey,* vol 6. Leicester: British Psychological Society.

Eysenk, M. W. (1984). *A Handbook of Cognitive Psychology.* London: Lawrence Erlbaum Associates.

Fodor, J. A. (1983). *The Modularity of Mind.* Cambridge MA: Bradford Books/MIT Press.

Fodor, J. A., Bever, T. G. and Garrett, M. F. (1974). *The Psychology of Language.* New York: McGraw-Hill.

Galotti, K. M. and Ganong, W. F. (1985). What non-programmers know about programming: natural language procedure specification. *International Journal of Man-Machine Studies,* **22**, 1-10.

Garnham, A. (1985). *Psycholinguistics: Central Topics.* London: Methuen.

Garnham, A. (1987). *Mental Models as Representations of Discourse and Text.* Chichester: Ellis Horwood.

Gentner, D. and Stevens, A. L. (1983). *Mental Models.* Hillsdale, NJ: Lawrence Erlbaum Associates.

Gick, M. L. and Holyoak, K. J. (1980). Analogical problem solving. *Cognitive Psychology,* **12**, 306-355.

Gick, M. L. and Holyoak, K. J. (1983). Schema induction and analogical transfer. *Cognitive Psychology,* **15**, 1-38.

Goldberg, A. and Robson, D. (1986). *Smalltalk 80 – The Language and its Implementation.* Addison-Wesley.

Gross, T. F. (1985). *Cognitive Development.* California: Wadsworth.

Johnson-Laird, P. N. (1983). *Mental models.* London: Cambridge University Press.

Kahney, H. (1986). *Problem Solving: A Cognitive Approach.* Milton Keynes: Open University Press.

Kintsch, W. and van Dijk, T. A. (1978). Toward a model of text comprehension and production. *Psychological Review*, **85**, 363-394.

Kowalski, R. (1982). Logic programming for the fifth generation. *In* K. L. Clark and S.-A. Tärnlund (Eds), *Logic Programming.* Amsterdam: North Holland.

Larkin, J. (1981). Enriching formal knowledge: A model for learning to solve textbook physics problems. *In* J. R. Anderson (Ed.), *Cognitive Skills and Their Acquisition.* Hillsdale, NJ: Lawrence Erlbaum Associates.

McKeithen, K. B., Reitman, J. S., Rueter, H. H. and Hirtle, S. C. (1981). Knowledge organization and skill differences in computer programmers. *Canadian Journal of Psychology,* **13**, 307-325.

McKendree, J. and Anderson, J. R. (1987). Effect of practice on knowledge and use of basic Lisp. *In* J. M. Carroll (Ed.), *Interfacing Thought: Cognitive Aspects of Human-Computer Interaction.* Cambridge, MA: MIT Press.

Mani, K. and Johnson-Laird, P. N. (1982). The mental representation of spatial descriptions. *Memory and Cognition,* **10**, 477-488.

Manktelow, K. I. and Jones, J. (1987). Principles from the psychology of thinking and mental models. *In* M. M. Gardiner and B. Christie (Eds), *Applying Psychology to User-Interface Design.* Chichester: Wiley.

Marr, D. (1982). *Vision: A Computational Investigation in the Human Representation of Visual Information.* San Fransisco: Freeman.

Miller, G. A. (1956). The magical number seven, plus or minus two: some limits on our capacity for information processing. *Psychological Review,* **63**, 81-97.

Miller, L. A. (1974). Programming by non-programmers. *International Journal of Man-Machine Studies,* **6**, 237-260.

Minsky, M. (1975). A framework for representing knowledge. *In* P. H. Winston (Ed.), *The Psychology of Computer Vision.* New York: McGraw-Hill.

Newell, A. and Simon, H. A. (1972). *Human Problem Solving.* Englewood Cliffs: Prentice-Hall.

Ormerod, T. C., Manktelow, K. I., Robson, E. H., and Steward, A. P. (1986). Content and representation effects with reasoning tasks in Prolog form. *Behaviour and Information Technology,* **5(2)**, 157-168.

Ormerod, T. C., Manktelow, K. I., Steward, A. P. and Robson, E. H. (1990). The effects of content and representation on the transfer of Prolog reasoning skills. *In* K.J. Gilhooly, M.T.G. Keane, R.H. Logie and G. Erdos (Eds), *Lines of Thinking.* Chichester: Wiley.

Paivio, A. (1971). *Imagery and Verbal Processes.* New York: Holt, Rinehart and Winston.

Pylyshyn, Z. W. (1979). Imagery theory: Not mysterious – just wrong. *The Behavioral and Brain Sciences,* **2**, 561-563.

Pylyshyn, Z. W. (1984). *Computation and Cognition.* Cambridge, MA: Bradford Books/MIT Press.

Reason, J. T. (1979). Actions not as planned: the price of automatization. *In* G. Underwood and R. Stevens (Eds), *Aspects of Consciousness.* London: Academic Press.

Reed, S. K., Dempster, A. and Ettinger, M. (1985). Usefulness of analogous solutions for solving algebra word problems. *Journal of Experimental Psychology: Learning, Memory, and Cognition,* **11(1)**, 106-125.

Rist, R. S. (1986). Plans in programming: definition, demonstration, and development. *In* E. Soloway and S. Iyengar (Eds.), *Empirical Studies of Programmers.* Norwood, NJ: Ablex.

Rumelhart, D. E. and McClelland, J. L. (Eds). (1986). *Parallel Distributed Processing: Explorations in the Microstructure of Cognition,* vol. 1: Foundations. Cambridge, MA: Bradford Books/MIT Press.

Rumelhart, D. E. and Norman, D. A. (1983). Representation of knowledge. Technical Report No. 116, Center for Human Information Processing, San Diego, University of California. Reprinted in Aitkenhead, A. M. and Slack, J. M. (Eds) (1985). *Issues in Cognitive Modelling.* London: Lawrence Erlbaum Associates.

Schank, R. C. (1972). Conceptual dependency: a theory of natural language understanding. *Cognitive Psychology,* **3**, 552-631.

Schank, R. C. (1982). *Dynamic Memory.* New York: Cambridge University Press.

Schank, R. C. and Abelson, R. P. (1977). *Scripts, Plans, Goals, and Understanding.* Hillsdale, NJ: Lawrence Erlbaum Associates.

Searle, J. (1980). Minds, brains, and programs. *The Behavioral and Brain Sciences,* **3**, 417-457.

Shiffrin, R. and Schneider, W. (1977). Controlled and automatic human information processing: II. Perceptual learning, automatic attending, and a general theory. *Psychological Review,* **84**, 127-190.

Sime, M. E., Arblaster, A. T. and Green, T. R. G. (1977). Reducing errors in programming conditionals by prescribing a writing procedure. *International Journal of Man-machine Studies,* **9**, 119-126.

Simon, H. A. (1978). Information processing theories of human problem solving. *In* W. K. Estes (Ed.), *Handbook of Learning and Cognitive Processes.* Hillsdale, NJ: Lawrence Erlbaum Associates.

Soloway, E. and Erlich, K. (1984). Empirical studies of programming knowledge. *IEEE Transactions on Software Engineering,* **10(5)**, 595-609.

Tulving, E. (1985). How many memory systems are there? *American Psychologist,* **40**, 385-398.

Wason, P. C. (1966). Reasoning. *In* B. M. Foss (Ed.), *New Horizons in Psychology,* vol. 1, Harmondsworth: Penguin.

Weiser, M. and Shertz, J. (1983). Programming problem representation in novice and expert programmers. *International Journal of Man-Machine Studies,* **19**, 391-398.

White, R. (1988). Effects of Pascal knowledge on novice PROLOG programmers. DAI research paper No. 399, Department of Artificial Intelligence, University of Edinburgh.

Wilensky, R. (1981). Meta-planning: Representing and using knowledge about planning in problem solving and natural language understanding. *Cognitive Science,* **5**, 197-233.

Chapter 1.5

Methodological Issues in the Study of Programming

David J. Gilmore

Psychology Department, University of Nottingham, Nottingham, NG7 2RD, UK

Abstract

The primary purpose of this chapter is to provide readers with a sufficient understanding of methodological issues to enable them to consider critically the numerous experimental results presented elsewhere in the volume, particularly given the potential conflicts between controlled laboratory studies and real-world observations. The chapter commences with a review of the many reasons why data might be collected in the study of programming, before looking at the choice of programming tasks for investigation. This is followed by an introduction to the key concepts of experimental design – statistical significance, effect size, sample size and power – since understanding the distinctions between these is critical in applied research, even when behaviour is being observed in real-world contexts. Methods for observing programming behaviour *in situ* are presented next, followed by a brief discussion of some special issues which arise when we try to generalize from applied research. The chapter concludes with some example case studies which are intended to emphasize the complementary nature of controlled methods versus observation and artificiality versus the real world.

Psychology of Programming
ISBN 0-12-350772-3

1 Types of data collection

When planning or assessing research it is important to consider the reasons why the research is to be, or was, conducted. This is a topic which I shall return to later, but it is necessary here to outline four of the main reasons why data is collected. It would be a mistake to believe the distinctions are as clear as I present them here, particularly since research may be conducted for more than one reason.

1.1 Hypothesis testing

This is the main tradition in cognitive psychology and it is used when contrasting theoretical positions make different predictions about the effect of some manipulation on behaviour. In these circumstances it is desirable to set up experimental situations with everything constant, except for the essential manipulation. In this way any observed difference in performance can be attributed to that one difference.

For example, one could have a theory of program comprehension that claims comprehension is attained through an initial analysis of syntactic structure and, therefore, that the use of indentation to indicate syntactic structure will lead to improvements in all aspects of program comprehension. An alternative theory might simply be that much program comprehension is possible without an analysis of syntax. This theory predicts benefits for indentation only when the programming task requires understanding of the program's syntactic structure. In this, rather simple, scenario an ideal hypothesis-testing experiment can be constructed. Subjects can be supplied with indented, or non-indented programs and then be required to perform tasks which may or may not require comprehension of the syntactic structure.

Unfortunately life is not always as simple as in this example. Frequently it is not possible to make one manipulation without 'knock-on' effects resulting in experimental conditions that differ in more ways than the hypothesis requires. For instance, in the above example, indentation might mean that long lines must be wrapped onto the next line. In this case, the experimental design is impure, since any effect observed may be due either to indentation or to line wrapping, or the two effects may cancel each other out. In these circumstances the researcher must introduce new conditions that may not seem realistic, but which allow the hypothesized effect to be properly observed. In the indentation example one would have to compare indentation with and without line wrap, and line wrap with and without indentation.

Assessing the outcome of such experiments relies on statistical tests to determine the probability that the observed differences are due to chance variations in performance (assuming there is no difference in reality), rather than systematic variations. This probability is described as the statistical significance of the effect (see below).

A vital component in any hypothesis testing research is the theoretical framework, since this both enables the researcher to make decisions about what tasks to provide and what data to measure and it provides the context within which the results can be explained. Hypothesis-testing research is not possible without a theoretical framework and such research is better described as a comparison.

1.2 Comparisons

Similar to hypothesis testing, but less systematic, are comparisons. In these the researcher attempts to discover which of two (or more) alternatives is easier to use, or whether some change to the programming process has an effect on performance. Comparisons are different from hypothesis testing in that they are intended to observe, but not explain any effect. Whereas hypothesis testing is used mainly in theoretical developments, comparisons are commonly used to recommend courses of action. Although comparisons cannot provide explanations, since the cases to be compared often differ on many criteria simultaneously, they are excellent at stimulating hypotheses and theoretical frameworks.

To return to our indentation example, our research question may simply be whether indentation provides a more usable representation of a computer program. For this comparison, although the results themselves may differ according to the presence or absence of line wrap in the actual experimental materials, line wrap is not important in interpreting the results, since it is a necessary feature of the indentation of real programs.

The results of such comparisons are conventionally tested for statistical significance, even though the more useful measure would be one of the size of the difference between the two conditions. This is known as the effect size (for more details, see below). The effect size is critical when results lead to recommendations for action, since there may be hidden costs in the action that were not assessed in the comparison (e.g. time spent producing flowcharts, printing facilities for flowcharts, etc.), but which need to be taken into account before action is taken.

1.3 Evaluations

The distinction between evaluations and comparisons is a very fine one and there is considerable overlap. However, an evaluation occurs when we are asking a question like 'Can people use flowcharts?', rather than 'Are flowcharts better than structure diagrams?'. Evaluations may also be performed simply with the intention of improving a system. Thus, it is unnecessary to compare it with something; rather, one looks for the weaknesses in the system. Whereas with comparisons we may design our experiment with only one or two measures of performance, in evaluations it is common to measure as much as possible. Subjective, non-performance measures (i.e. questionnaires) are usually used as well, since preferences may be as important as performance. Decisions as to which of the many measures are most informative will often be made *post hoc*. For example, in evaluating a novice programming environment we may be interested in showing that 50% of programs run first time or that 50% of bugs are corrected within 5 minutes. These criteria may have been derived from observations of other systems, but note that no direct comparison occurs.

1.4 Exploration

Exploratory data collection occurs when the research question begins 'What happens if ...?', or 'How do people do ...?'. Such questions are particularly common when the main research theme is novel and we lack enough information to engage in the other types of data collection (as when studying new paradigms, such as object-oriented, parallel or logic programming). For example, if we are interested in teaching aids for some programming language we may decide to collect examples of programs written by novices, in order to look at the sorts of errors they make, how they correct them, etc. Alternatively we may wish to look at problems that arise in system design generally, rather than just in the coding stage, which may involve observations of a number of examples of system design.

The prime difficulties with exploratory data collection are that it can be very difficult to collect and record such data and it is rare for the resultant data to be well defined. Thus, we can try to count novice programming bugs, but it will be difficult to decide what counts as a single bug, and the number produced will not help us to describe the errors. Likewise there are no obvious numbers to be collected from observations of the system design process. Although numbers are not essential to all research, the description and interpretation of non-numeric data can be very difficult.

1.5 Summary

The above methods of data collection are each appropriate in different circumstances. The results of each need to be interpreted differently, and the results from each are often necessary for the design of further research using other approaches.

There is no simple relationship between the ease of data collection and result interpretation. Thus, hypothesis-testing research often requires much work in the construction of experimental conditions and materials. However, if designed correctly the results should be directly interpretable in terms of the theoretical justification of the experiment. Conversely, although exploratory data is often easier to collect, its analysis can be extremely time consuming.

2 The experimental task

As already noted, there are many different kinds of research goals, and as well as requiring different methods, they are also best examined through different tasks. Exploratory data collection will require the use of realistic programming tasks, since the main goal is to observe real programming activity. In contrast, the hypothesis-testing approach is not bound by the need for realism and can construct experimental tasks better suited to testing the hypothesis under consideration.

Many different tasks have been given to programming subjects. The more obvious include asking subjects to write a program and 'think aloud' while they do it, or giving them a program and setting them a comprehension quiz about it. Another very natural technique is to scrutinize code, often from students, either to classify types of errors (Spohrer and Soloway, 1986) or to compare frequencies of different solution types (Ratcliff and Siddiqi, 1985). But there are also less obvious, more artificial tasks, such as timed question answering with miniature programming languages (Sime *et al.*, 1977), or recalling progams that have only been viewed briefly (Kahney, 1983) Each task has its place. In this section I shall illustrate some of the reasons for choosing different types of tasks.

There are two important considerations – the object to be studied (the person or the system), and the programming task of primary interest (see Chapter 1.3). I shall examine these in turn, but the same conclusion can be drawn from each: 'You can't study people using one system and assume your results generalize; you can't study a single task and assume you have found out all the important facts'.

The prime interest in studying the person has been in the comparison of experts and novices, with particular interest in how experts represent programming knowledge. The most popular approach has been exploratory techniques, such as verbal protocols and recall techniques (revealing what is recalled first, what is forgotten, etc.). In both, the results depend upon the investigator's judgement to extract meaningful patterns from the data. Subsequent testing to discover whether the patterns are real or spurious requires more rigorous approaches based on comparisons and hypothesis testing, such as the priming study by Pennington (1987) outlined below, or 'fill-in-the-blank' (Cloze) tests, in which we predict the ease with which different programmers will be able to complete different types of program statements.

Studying the system, unlike studying the person, does not seem to invite exploratory study. Typical approaches have been the comparison of different notational structures (Sime *et al.*, 1977) and studies of different browser designs (Monk and Walsh, 1988). These are experiments where performance was compared in two or more different experimental conditions, using very orthodox laboratory designs. The tasks used in such studies usually cover a range of programming activities. For example, Sime *et al.* used carefully timed question answering and coding, measuring time, number of syntactic errors, correction time, etc. The use of miniature programming languages, novice subjects and a game-like scenario enabled accurate conclusions to be drawn about notational differences, though at the expense of reality.

Studies of different programming tasks (i.e. writing, reading, debugging, modifying, etc.) offer another perspective on the choice of experimental tasks. The majority of studies of program writing have used observational techniques, either video (with or without verbal protocols) or *post hoc* analysis of code. Exceptions include the work of Hoc (1981) comparing different environments for solving data-driven and procedure-driven problems, and the work of Sime *et al.* on writing programs in different notations. Studies of program comprehension, in contrast, have used a great variety of different techniques. A common technique is a comprehension quiz, but this is not as simple as it might seem. Some studies have failed to detect effects of their manipulations (e.g. indentation) because when all the test items are averaged together there is no effect, but there may be large differences on a few of the items. The challenge to the researcher is then to explain why those few items are different. Other comprehension techniques include hand-execution, 'fill-in-the-blank' tests (Soloway and Ehrlich, 1984; Thomas and Zweben, 1986) and the recognition of program structures (Cunniff and Taylor, 1987). Studies of debugging have used both observational techniques and also comparisons of performance with and without various tools (Weiser, 1982, 1986; see below).

Here again there is an important warning: whilst single tasks, single subject populations (e.g. students) and single languages may be used to disprove a theoretical hypothesis, they are of little use when they confirm hypotheses. Early studies of programming languages often used a single task, such as hand-execution, to compare different notations. However, Green (1977) showed that different procedural notations indistinguishable on one task (reasoning about control-flow), could produce very different performance on a second task (reasoning about conditional structures). Later, Gilmore and Green (1984) were successful in showing that the effect partially reversed with declarative languages. Reasoning about control-flow revealed notational differences, where reasoning about conditional structures did not (more details are given in Chapter 2.2).

In summary, therefore, for truly generalizable conclusions, a variety of tasks, a variety of types of programmers and a variety of languages should be studied. When all these variations are not present, then care must be taken not to overgeneralize the results.

3 Important concepts in experimental design

3.1 Statistical significance

This concept is probably familiar to readers, since it is the most frequently reported in published papers. It provides a measure of the likelihood of the observed difference in scores being a result of chance variations in the performance of four subject groups, assuming that they were drawn from the same population and that the experimental manipulations had no influence on their performance. As researchers we are interested in drawing the conclusion that our subject groups, although from the same population originally, have been differentially affected by the manipulation and can now be regarded as being from different populations. Thus, the statistical significance of a result assesses the probability that there is not a difference in the scores achieved by the different groups. Only when this probability is small can we draw the inference that our experimental manipulation had an effect on performance.

By convention a probability of 5% is taken as this small level beyond which we conclude that we have observed a real effect. Thus an observed difference with a 6% significance is commonly not reported, where a 4% probability might lead to publication. Because statistical significance is simply a measure of probability, we should take care to interpret it as such. A probability of 5% represents a one in twenty chance, which means that if we conduct twenty statistical tests, then there is a high probability that one will reveal a significance of around 5%, without any implication that our samples are from different populations. This can be a substantial problem when gathering large quantities of data, since this may require numerous statistical tests. The best solution is to use hypotheses to restrict the number of tests performed and to use significance levels smaller than 5%.

3.2 Effect size

A concept which is frequently confused with statistical significance and which is of the utmost importance when interpreting applied research is that of effect size. This reflects the influence on a score of the experimental manipulations, though not just by magnitude, but also by variability. Thus effects which are easiest to detect are either small in magnitude and totally reliable across subjects, or more variable, but large in magnitude. Much harder effects to detect are those which have both small magnitude and large variability.

Unfortunately effect size is not usually a calculable quantity, though estimates can sometimes be made by statistical analysis (see Hoc and Leplat, 1983, for further details). Nevertheless, the larger the effect the easier it will be to observe in an experiment. In such circumstances fewer subjects may be needed, or noisy measures of performance[1] may be used.

A statistically significant effect is not equivalent to an ecologically relevant effect, since the effect may be small, but the experimental precision sufficiently high that

[1] A noisy measure is one in which the task context or the mechanism of measurement introduces extra variability into the data (e.g. a task context in which many different strategies are available to subjects).

it is detected statistically. Correspondingly, a non-significant effect is not equivalent to an ecologically negligable effect, since the experimental precision may have been so poor that the effect could not be detected statistically. This second point is of great importance, since it is quite common for substantive conclusions to be based on the flimsy fact that the results were not statistically significant. Landauer (1987) provides further discussion on this topic in relation to HCI generally, along with numerous examples.

3.3 Sample size

The problem of sample size is one of the most difficult to deal with, since it seems abundantly clear that small samples must be a bad thing. This intuition is usually based on the problem of the representativeness of the sample, which is determined by the method used to select the sample.

However, if the sampling method is unbiased, then small samples are not inherently bad. Indeed, the same levels of statistical significance from small and large samples would be suggestive of larger- and smaller-effect sizes, respectively. If the same experiment were conducted with either twenty or forty subjects and the same average scores and significance levels were obtained, then contrary to intuition the experiment with twenty subjects represents a larger and probably more ecologically important effect.

The assessment of statistical significance is based on the sample size, and it is incorrect, therefore, to talk of a result being 'more significant' because of the sample size (whether small or large).

The main problem with small samples arises when the results are not significant (i.e. $p > 5\%$), since the use of a larger sample, all else being equal, might reduce p to below 5%. Thus again, one must be cautious when interpreting non-significant results, especially when the statistical significance is in the 5% to 20% region[2].

It is important to note also that the sample size is not the total number of subjects used, but the number used in each experimental condition. The choice of sample size is frequently constrained (by availability, expense, etc.), preventing the use of an adequate number for the anticipated effect size. When this situation arises, it is necessary to increase the power of the experiment through more careful design.

3.4 Power

The power of an experiment is a tricky concept. It is a measure of the quality of an experimental design. It is affected by sample size and by the quality of the measures of performance, the accuracy of the classification of subjects into groups and by the quality of the materials used.

Imagine we are investigating the benefits of a new debugging tool. The effect size is constant but, according to the experimental design, our ability is reflected by the accuracy of these estimates. Power will be affected by the quality of our performance measures, not just in terms of millisecond accuracy, but also in terms of how well defined and reliable across subjects our measures are. We could also use either a

[2]In general, it seems to be the case that the sensitivity of an experiment is proportional to the square root of the number of subjects. Thus to double the sensitivity of an experiment around four times as many subjects will be required. Adding a few extra subjects to the sample is unlikely to make much difference.

within-subjects design (where each subject performs in each experimental condition) or a between-subjects design (where each subject performs in only one experimental condition). In general a within-subjects design is more powerful, producing the same level of significance with fewer subjects. However, the repetition in a within-subjects design (which may lead to learning and order effects) may lessen the quality of the performance measures, thereby decreasing the power.

The main skill of experimental design is constructing an experiment of sufficient power to detect an effect, without overkill. When contradictory results are discovered (one with no significant differences and one with significant differences) it is important to remember that they may be due to differences in the power of the experimental designs.

3.5 Summary

(1) Statistical significance measures probability and is not a direct measure of effect size.

(2) Significance levels reflect both effect size and the power of the experimental design.

(3) Large effects require less-powerful experimental designs.

(4) Small sample sizes produce less-powerful designs.

(5) Noisy measures of performance produce less-powerful designs.

(6) Weak designs may fail to reveal genuine, but small, effects.

4 Observational techniques

Within the life sciences generally, use is made of both observational and experimental techniques. The former serve in the process of discovery when existing theoretical frameworks are unable to precisely define hypotheses, whilst the latter can validate hypotheses. The former can be particularly valuable for uncovering new and interesting phenomena, but they are unable to fully explain them. The latter will produce well-defined information, but are unlikely to discover something new. Observational techniques are a form of exploratory data collection, but they are becoming increasingly sophisticated and, thus, deserve a separate section of this chapter.

In the study of programming it may seem surprising that initial research focused on experiments, when observational methods would seem more appropriate to a new discipline. However, the first investigators either performed simple comparisons or developed their hypotheses from existing theoretical frameworks in psychology. The shift to observational techniques has grown from doubts about both the value of generalizing from context-bound comparisons and the further applicability of the theoretical frameworks.

Green (1980) provided a good review of some of this early research, covering three theoretical frameworks: 'Languages as Notations', 'Programming and Natural Languages' and 'Programming as Problem Solving'. The integration of these frameworks is the challenge to future research, requiring observational studies to stimulate the development of new unifying ideas.

Many of these observational techniques involve verbalization by programmers about either their current activity or, retrospectively, about some earlier activity. This is known as a protocol, and the analysis of such data is known as protocol analysis. Due to problems associated with the reliability of people's insight into their own thought processes, there are many complex issues in this area. Ericsson and Simon (1984) provide a clear and comprehensive coverage of these issues and is essential reading for any researcher contemplating the use of protocol analysis.

There are a number of different areas where such protocols might be collected:

(1) Interviews with programmers: Prior to the use of more formal techniques a researcher may conduct an interview with a programmer about the nature and organization of the work. The information acquired should not be assumed correct since there may be a discrepancy between actual activity and the planned, intended activity as perceived by the programmer. The more formal observations will provide an opportunity to verify these initial findings.

(2) Video plus verbalization: Video provides a relatively easy means of observing the programmer's 'natural' programming activities. These may be accompanied by simultaneous verbalizations, or the programmer may review their performance later on. The former may interfere with the subject's normal programming style (through increased mental workload), whilst the latter may lead to a rationalized account of unstructured behaviour. The advantage of such methods is that, unlike simple error counts, etc., they are able to capture the full dynamic nature of programming activity.

(3) Constructive interaction: A more natural form of observation may be to have two users co-operating on the solution to a problem. In this way their conversation provides natural verbalization. The main problem with this method is that of making an individualistic interpretation of the results. This method has not been used in relation to programming problems, though examples of its use can be found in O'Malley *et al.* (1984).

(4) Longitudinal studies of learning: Over the length of a programming course attitudes and perceptions can be assessed and submitted programs collected. The sheer volume of data that this produces is a problem in its own right, with any analysis being extremely time consuming and frequently subjective. These methods are further complicated by the uncontrollability of the situation. Prior experience, books read, experiments performed, effort expended are all uncontrolled variables which may influence the learning process.[3]

(5) Social and organizational processes: The larger context of programming (in a commercial/industrial world) is affected by many social and organizational issues. These cannot really be studied using traditional hypothesis-testing techniques and require observational methods. Chapter 4.1 provides an example of how such research may be conducted.

[3]In fact, these variables affect all studies of programming, though hypothesis-testing methods can control for them, in part, through careful experimental design. The only controls available for observational techniques are *post hoc* statistical ones (i.e. multiple linear regression). However, the increasing availability of easy-to-use statistical packages means that such controls may become more powerful than experimental ones.

In fact, it is misleading to pretend that there is a clear distinction between hypothesis-testing and observational techniques, since there are increasing number of studies that use aspects of both. For example, Widowski (1987) uses a simulation of a programming environment to capture elements of natural behaviour, but within a controlled environment. Leventhal and Instone (1989) describe a pilot longitudinal study in which formal experimental techniques are used at regular intervals throughout a programming course. This mixed research is likely to prove the most fruitful over the next few years.

5 Applying research

5.1 What was the question?

In the hypothesis-testing approach the question addressed by a piece of research is usually straightforward and clearly presented. Thus, for example, we might ask the question 'can we demonstrate circumstances in which novice behaviour is superior to that of experts?' (Adelson, 1984). But it is rare that we have such precise questions to answer. This is usually more of a problem for an assessor of the research than for the researcher. It is not unusual for research to be criticized for failing to consider certain factors or issues, when they were not relevant to the researcher's own question.

For example, the research conducted at the University of Sheffield in the 1970s (Sime *et al.*, 1977) into the design and use of different conditional structures (e.g. If ... GOTO versus IF ... THEN ... ELSE ...) was intended to further our understanding of how program representation interacts with the programming task to determine performance. Although it was not intended that the results be taken as strict recommendations for future language design, Sheil (1981) has dismissed these research efforts with

> 'how compelling does one find a psychological study of novice programming which found novices' fears of dealing with a teletype so overwhelmed their effects that they studied 'programming' in the context of a bizarre game which involved feeding a mechanical rabbit!'

The statistical argument behind Sheil's criticisms is that through using experimental design with high power, Sime *et al.* have detected effects of small size, which are of no practical value, an argument that is valid within the comparison approach to research, but not within hypothesis-testing, for which the design of Sime *et al.* was well suited.

Thus, hypothesis-testing research may not be directly useful for the production of human factors guidelines, though the resultant clarification of theory should be.

5.2 Interpreting causes

Conversely, in the human factors paradigm it is often sufficient just to show, for example, a 30% improvement in performance with system X as opposed to system Y. In these situations one must be wary of trying to draw theoretical conclusions from the design difference between X and Y since such inferences are often not valid. For example, we may compare experts in different programming languages, but any differences cannot then be attributed simply to language differences since the programmers may also differ in their programming education, motivation and programming problems experienced. Gilmore and Green (1988) found marked differences in the nature of debugging by BASIC and Pascal programmers.

Their language-specific conclusions were based on the fact that the subjects in these different languages had apparently had similar computing experiences and that the subjects performed comparably overall. But Davies (1988) has shown that BASIC programmers can perform like Pascal programmers when they have received tuition about program design skills. Given the difficulties of finding Pascal programmers who have not received such training, there has been no investigation of whether Pascal programmers without design skills perform like BASIC programmers. For now, we can only conclude that there are education effects and probably language differences too.

Alternatively, a study may compare novices and experts in the same language, gaining considerable knowledge about their differences, but without understanding the causes of the differences. This is especially important when we use our knowledge of novice-expert differences to recommend particular teaching practices. The critical factor may be amount of programming experience, or range of problems studied or the numbers of programming languages known.

This problem of understanding the cause of an observed effect is closely related to the problem of experimental validity, which is of great importance in assessing the value of any piece of research.

6 Experimental validity

The problem of validity arises when we try to generalize from our results to other samples, to other languages, or to theoretical positions. There are many different types of validity and it is important not to confuse them.

Two important types of validity are internal and external validity. Internal validity describes our ability to be sure that our explanation of the observed differences is the only likely explanation, and it is closely related to the statistical issues discussed earlier. External validity describes our ability to make correct generalizations about the implications of the results. In most cases external validity is only an issue when we have high internal validity. Hypothesis-testing research must place most emphasis on internal validity, while comparisons and evaluations need to emphasize external validity.

For example, the Sime *et al.* studies (and the later Gilmore and Green studies mentioned above) have high internal validity.

The differences observed cannot be due to anything other than the differences in the notations presented to subjects. These studies demonstrated that the subjects needed different types of information for different tasks and that an important property of the notations used was the match between the information easily obtained and that needed for the task.

These studies have moderate external validity, since the result cannot be unquestionably generalized to all programming notations and programming tasks (a result that would be of great interest to anyone developing new notations). The results should generalize, however, to novice programmers learning a first language and probably to novices learning any language. Since the subjects in the experiments did not become experts in the notations, it is hard to be sure of the external validity in relation to expert programmers.

Although not lacking in external validity, these studies do lack face validity, a concept that is easily confused with external validity. Whereas external validity reflects our actual ability to generalize, face validity reflects our apparent ability to generalize.

Face validity can be illustrated through one of Sheil's criticisms of these experiments. Sheil appeals to his readers' intuitions about the compelling nature of research involving a 'mechanical rabbit' without discussing whether the important conclusions depend upon the method used. Given that the main implications of the Sime *et al.* studies derive from the theoretical position, the method of study is of little relevance.

Unfortunately, research can have high face validity without either internal and external validity. For example, Gannon (1977) assessed the value of data typing in programming languages. To do this, he developed two complete languages that were identical in all respects except data typing and, unavoidably, any other features affected by data typing. Although his results show an advantage for statically typed languages, it is not clear whether this is solely due to data typing, or whether the advantage derives from the 'better primitives with which to solve string-processing problems', or the fact that 'most students learn a statically typed language as their first language'. Any explanation of these results cannot be shown to be the only valid explanation (an absence of internal validity). Without internal validity there is unlikely to be a high level of external validity.

Face validity is an important property of research, especially when there is a need to convince others of the value of the research. However, it should be considered *after* internal and external validity, not before. Programming behaviour is extremely complex and in order to acquire statistically useful measures of performance simple tasks and unreal measures of performance are often essential. This is particularly so when we wish to make inferences about the causes of observed differences in performance.

7 Methodological case studies

7.1 Current concerns rigorously applied

There is a considerable literature on the nature of expertise in programming (see Chapter 3.1), but very little of it has used the hypothesis-testing approach to validate the many observational results. Pennington (1987) carried out a complex study that took a standard psychological methodology and applied it to programming. Although the study has little or no face validity, it has high internal validity, through a tradition in cognitive psychology (McKoon and Ratcliff, 1980). Pennington's experiment addressed the expert programmer's mental representation of a computer program. The subjects' task was to comprehend and learn a program.

Pennington tested the programmer's memory for particular lines of the program through a recognition test. However, before presenting the statements to be recognized, she presented subjects with a momentary glimpse of another statement from the program. If the mental representation of this 'priming' statement is close to the 'to be recognized' statement then the speed with which recognition occurs will be quicker. If there is no mental association between prime and test statements then the recognition time will be slower.

By using primes that were either procedurally or functionally connected to the test item it is possible to infer the mental distance between statements in the program.

Pennington's results revealed that

> procedural rather than functional units form the basis of expert programmers' mental representations. (Pennington, 1987, p. 295)

This result, which is different from others in the area, is important because of the high internal validity. It serves not to further applications of existing results, but to qualify them and point out possible errors of interpretation in earlier experiments. This reveals the value in using the hypothesis-testing approach for theories that have been derived from comparative and evaluative data.

7.2 The real world raises problems of representation

At the opposite end of the spectrum we can consider a valuable observational study with high face validity, but which does not lend itself to generalizable conclusions. Letovsky *et al.* (1987) looked at an IBM code inspection process. They analysed one sixty-five minute video tape in detail and found that three goals were being achieved during the inspection (clarity, correctness and consistency) through the use of three behaviours (reconstruction, simulation and cross-checking). A feature of their analysis was that although they were observing a high-level process, the problems encountered during the code inspection varied from syntactic concerns to problems of specification. Of particular interest are the syntactic difficulties which arose with the use of 'case-statement' in which there was conflict between the 'natural' task structure and the most efficient code structure. The design team had to make decisions about balancing comprehensibility and efficiency.

Although it is not clear how an alternative syntax could have solved this problem, it illustrates that the mapping from syntax to semantics can be a critical part of the programming process. Thus, unlike a hypothesis-testing experimental study, which tends to look for one clear-cut result, observational studies can provide interesting results along a variety of dimensions.

7.3 The challenge to external validity

Weiser (1982) investigated aspects of expert debugging strategies and found that experts use 'slices' when debugging. A 'slice' is that part of the program which can alter the value of a particular variable (the buggy variable). An important property of a slice is that the statements within it need not be textually contiguous, but may be scattered throughout the program.

In his experiment, Weiser's subjects had to debug Algol-W programs of seventy-five to 150 lines, which contained just one error in the subroutine which performed the bulk of the calculations. To test the programmers' use of slices, they were given a recognition memory test after debugging all three programs. The test examined their memory for slices versus contiguous code and for relevant (to the bug) and irrelevant statements. Since the interest was in semantic understanding, details such as variable names were altered before the lines were presented for recognition.

The results of the experiment revealed that relevant slices were recognized as accurately as relevant contiguous statements, but significantly better than the irrelevant slices. From this Weiser concludes that although programmers do not use only slices when debugging; this is clear evidence that slices are used as part of the debugging strategy.

Given that slices can be formally defined, the natural progression of this research was to build slice-based debugging tools. Weiser and Lyle (1986) evaluated the utility of such tools for debugging. Despite the apparent validity of the early experiments, these evaluations revealed that although 'pencil and paper' simulations of slicing aids were beneficial in speeding up debugging, the on-line tool showed no such benefits. Weiser and Lyle conclude that

> perhaps, slicing is like watching a beautiful sunset – a computer can do it, but it just isn't the same. (p. 189)

Weiser and Lyle went on to look at dicing tools, which reduce the size of a slice still further (slicing the slice), and found that these tools did provide the programmers with an advantage when debugging, even though there were no results to show that expert programmers use dice when debugging.

This example reveals the need to be cautious when assessing the external validity of some research. The research results might seem quite conclusive, but the step from them to application may be considerable.

8 Summary

This chapter has presented various important concepts in experimental research and demonstrated that the study of computer programming skills requires a wide range of approaches to the collection of data.

An understanding of hypothesis-testing techniques is essential, since almost all of the early research on programming has this form and, at least until there is an agreed theoretical framework, it will continue to be of great importance in studying programming. But recently, observational techniques have become popular due to their ability to capture the more complex aspects of programming. Such techniques are now an invaluable part of the empirical repertoire, even though there are still some analytical problems.

The case studies demonstrate how a variety of approaches, tasks and methods should be viewed not as a problem or a weakness, but as an essential property of a rapidly changing area which has lots of pressing, but very different questions (e.g. theoretical, practical, speculative). There are many more examples to be found and hopefully this chapter has helped to prepare the reader to appreciate those that are presented in the rest of this book.

References

Adelson, B. (1984). When novices surpass experts: the difficulty of a task may increase with expertise. *Journal of Experimental Psychology; Learning, Memory and Cognition,* **10**, 483-495.

Cunniff, N. and Taylor, R.P. (1987). Graphical versus textual representation: an empirical study of novices' program comprehension. *In* G. Olson, S. Sheppard and E. Soloway (Eds), *Empirical Studies of Programmers: Second Workshop.* Norwood, NJ: Ablex.

Davies, S. (1988). The nature and development of programming plans. Paper presented at *International Conference on Thinking,* Aberdeen, 1988.

Ericsson, K.A. and Simon, H. A. (1984). *Protocol Analysis: Verbal Reports as Data.* Cambridge, MA: MIT Press.

Gannon, J.D. (1977). An experimental evaluation of data-type conventions. *Communications of the ACM,* **20**, 584-595.

Gilmore, D.J. and Green, T.R.G. (1984). Comprehension and recall of miniature programming languages. *International Journal of Man-Machine Studies,* **21**, 31-48.

Gilmore, D.J. and Green, T.R.G. (1988). Programming plans and programming expertise. *Quarterly Journal of Experimental Psychology,* **40a**, 423-442.

Green, T.R.G. (1977). Conditional program statements and their comprehensibility to professional programmers. *Journal of Occupational Psychology,* **50**, 93 - 109.

Green, T.R.G. (1980). Programming as a cognitive activity. *In* H.T. Smith and T.R.G. Green (Eds), *Human Interaction With Computers.* London: Academic Press.

Hoc, J-M. (1981). Planning and direction of problem solving in structured programming: an empirical comparison between two methods. *International Journal of Man-Machine Studies,* **15**, 363-383.

Hoc, J-M. and Leplat, J. (1983). *International Journal of Man-Machine Studies,* **18**, 283-306.

Kahney, H. (1983). Problem solving by novice programmers. *In* T.R.G. Green, S.J. Payne, and G. van der Veer (Eds), *The Psychology of Computer Use.* London: Academic Press.

Landauer, T. (1987). Cognitive psychology and computer system design. *In* J. Carroll (Ed.), *Interfacing Thought.* Cambridge, MA: MIT Press.

Leventhal, L. and Instone, K. (1989). Becoming an expert: the process of acquiring expertise among highly novice computer scientists. Paper presented at *Psychology of Programming Interest Group: First Workshop.* Warwick University, January, 1989.

Letovsky, S., Pinto, J., Lambert, R. and Soloway, E. (1987). A cognitive analysis of a code inspection. *In* G. Olson, S. Sheppard and E. Soloway (Eds), *Empirical Studies of Programmers: Second Workshop.* Norwood, NJ: Ablex.

McKoon and Ratcliff (1980). Priming in item recognition: the organisation of propositions in memory for text. *Journal of Verbal Learning and Verbal Behaviour,* **19**, 369-386.

Monk, A.F. and Walsh, P. (1988). Browsers for literate programs: avoiding the cognitive overheads. *Proceedings of ECCE4, Fourth European Conference on Cognitive Ergonomics.* Cambridge, UK, September, 1988.

O'Malley, C. E., Draper, S. W. and Riley, M. S. (1984). Constructive interaction: a method for studying human-computer-human interaction. *In* B. Shackel (Ed.), *Proceedings of Interact 84.* Amsterdam: Elsevier North-Holland.

Pennington, N. (1987). Stimulus structures and mental representations in expert comprehension of computer programs. *Cognitive Psychology,* **19**, 295-341.

Ratcliff, B. and Siddiqi, J.I.A. (1985). An empirical investigation into problem decomposition strategies used in program design. *International Journal of Man-Machine Studies,* **22**, 77-90.

Sheil, B. (1981). Coping with complexity. Paper presented at Houston Symposium III: Information Technology in the 1980s.

Sime, M.E., Arblaster, A.T. and Green, T.R.G. (1977). Structuring the programmer's task. *Journal of Occupational Psychology,* **50**, 205-216.

Soloway, E. and Ehrlich, K. (1984). Empirical studies of programming knowledge. *IEEE Transactions on Software Engineering,* **5**, 595-609.

Spohrer, J.C. and Soloway, E. (1986). Analysing the high frequency bugs in novice programs. *In* E. Soloway and S. Iyengar (Eds), *Empirical Studies of Programmers.* Norwood, NJ: Ablex.

Thomas, M. and Zweben, S. (1986). The effect of program dependent and program independent deletions on software cloze tests. *In* E. Soloway and S. Iyengar (Eds), *Empirical Studies of Programmers.* Norwood, NJ: Ablex.

Weiser, M. (1982). Programmers use slices when debugging. *Communications of the ACM,* **25**, 446-452.

Weiser, M. and Lyle, J. (1986). Experiments on slicing-based debugging aids. *In* E. Soloway and S. Iyengar (Eds), *Empirical Studies of Programmers.* Norwood, NJ: Ablex.

Widowski, D. (1987). Reading, comprehending and recalling computer programs as a function of expertise. *Proceedings of CERCLE Workshop on Complex Learning.* Grange-over-Sands.

Part 2

Language Design and Acquisition of Programming

One important consequence of the increasing number of programming languages and the evolution of program design environments is the number of new issues they raise. As developed in the first part of this volume this evolution is due both to the necessity to solve new classes of problems and to have more powerful tools to support the programming activity. However, this evolution does not necessarily make the programming activity more easy to practice and to learn. Are some languages really better than others? If so, what for? What kind of languages are better for learning and/or for educational purposes? What kind of support environment is really efficient for professionals?

This section examines programming languages, their design, their use, their structural and semantic properties and their learning. It raises what are hoped to be relevant issues for language designers and educators through presentation of data on the ways in which experts and novices behave in programming. What is common to both is that no one programming language (or a programming environment) is a panacea. Rather they are tools that facilitate certain tasks by supporting expert and novice programming behaviour in an appropriate manner, while in some cases they may constitute an obstacle to efficient planning activities.

The first two chapters deal with language design. In Chapter 2.1 Petre discusses the distance separating language designers' and expert programmers' criteria of what constitutes a 'good' programming language. Key features differentiating them involves the interpretation of abstraction: designers prefer high-level expressions, whereas the experts feel the need to choose their level of abstraction and to be able to manipulate hardware when necessary. Designers tend to emphasize 'well-foundedness' and correctness, whereas expert programmers stress utility, control and efficiency. Chapter 2.2 by Green concentrates on the detail of language design and describes how notational aspects influence the process of programming. A program is viewed as an information structure. The programmer is the user of this structure who is called upon to accomplish different tasks: obtain information from a program, add new information, re-organize it, debug it, modify it, etc. The main argument is that the structure of information should match the structure of the task. A set of cognitive dimensions of notation is developed in order to analyse how a combination of a notation and the environment in which it is used affect usability. There is no universally good notation system, only adequate notations for specific tasks. These dimensions can be seen as initial guidelines to cognitive requirements in language design.

The next two chapters explore learning issues. The relationship between task structure and programming language semantics is discussed as a critical component of programming learning in Chapter 2.3 by Hoc and Nguyen-Xuan. Novices must learn not only new notations and new means of expression but also the operating rules of the processing device that underlies the language. It is argued that learning by doing and learning by analogy are privileged mechanisms of acquisition of these rules. The notion of the representation and processing system (RPS) is developed in order to identify and analyse the gap between novices' existing RPSs and those they have to construct and to suggest the training situations that facilitate the development of RPSs. The acquisition of operating rules is only the first step in programming learning: novices have to also learn programming concepts and structures. Chapter 2.4 by Rogalski and Samurçay examines the cognitive difficulties encountered by novices in learning the concepts and structures commonly presented in introductory programming classes. It is shown how dynamic mental models related to action execution act as precursors by playing both productive and reductionist roles, and become an obstacle when more static representations are needed. It is argued that training paradigms tend to place emphasis on the computational and procedural aspects of programming which prevents novices from learning problem modelling and programming as a function specification.

Chapter 2.5 by Mendelsohn, Green and Brna discusses issues related to the use of computers for general educational purposes. After examining the 'transfer of competence' and 'acquisition of new knowledge' hypotheses, the authors put forward an alternative proposition: the new representation and processing system that people acquire via programming may modify their strategy in analysing the objects on which they have programmed.

All the chapters in this part analyse current issues in programming learning and the cognitive requirements of programming activity in language design. The study of learning mechanisms is faced with unresolved theoretical and methodological problems. Observation of learning activities is generally too short to assess the real characteristics of activity—with obvious consequences on validity (the importance and duration of errors, etc.).

A number of questions still remain open. Although we know that individual differences are important factors affecting the learning process and performance in the use of specific programming languages and environments, very few concepts have been defined to characterize these differences systematically. Little work has been done as well on the retraining of professional programmers. How quickly and thoroughly can they acquire a new language or a design environment? How do previously acquired representations and procedures affect ease of acquisition of new knowledge?

Chapter 2.1

Expert Programmers and Programming Languages

Marian Petre

Instituut voor Perceptie Onderzoek/IPO, Postbus 513, 5600 MB Eindhoven, The Netherlands

Abstract

This chapter contrasts cursorily the aspirations of general-purpose programming language designers with some evidence about expert problem solving and programming behaviour. The contrast is summarized in a rough wish list of what experts want from general-purpose programming languages. The programmers' wish list differs from the aspirations of language designers less in detail than in emphasis: whereas the designers emphasize well-foundedness and correctness, the expert programmers emphasize utility, control and efficiency. It is argued that a programming language is a tool, not a panacea; tools make easy the tasks for which they are designed, but the outcome depends on the intention and expertise of the wielder.

Psychology of Programming
ISBN 0-12-350772-3

1 Introduction: the language-user/language-designer schism, and why practitioners complain about new languages

Programming languages, assessed as they tend to be in personal and mystical terms, attract a tenacious confusion of myth, of variable quality. The accumulation of programming language myth reflects the evolution of computing. Programming languages are artefacts developed in mixed environments: technological, social and philosophical. Assumptions – often unspoken, sometimes accidental – incorporated in those original environments may survive the environmental constraints that nurtured them and may become irrelevant, inappropriate, or even misguided when the environment changes. Evolution implies changes in truth; myths convenient or pertinent at their inception may obscure issues in their continuation.

Programming began in the hands of engineers and hackers who created improvements as needed to provide obvious advantages of speed and power. Other criteria for languages were personal. The point was to build a tool and exploit it; 'software' was a way to build up layers of tool enhancements.

Software has since moved into the hands of theoreticians whose criteria are different from the hackers'. Theoreticians intend to improve languages by design rather than just by demand and tinkering, and they aspire to make languages that conform to formal, usually classical, models (e.g. functional languages are based, with varying mathematical purity, on the lambda-calculus). This evolution produced a schism between those who use languages (engineers and hackers) and those who design them (theoreticians). There is a shift of orientation from utility to 'well-foundedness'.

Early programming languages were clumsy and occasionally dangerous because they were not well-founded. Modern languages often fail in practice because emphasis on theory displaces attention to use. This re-orientation is not inherently bad, although it can result in impractical languages. The re-orientation is, however, incomplete and hence misleading; critical decisions (e.g. choice of formal model, patterns of notation or implementation not covered by the model) remain ill-founded. For example, Kowalski champions logic programming on the disputable grounds that 'Symbolic logic was first designed as a formalization of natural language and human reasoning' and hence 'Logic reconciles the requirement that the...language be *natural and easy to use* with the advantage of its being machine-intelligible'. (Emphasis added; Kowalski, 1982.) Yet there is evidence to undermine these 'foundations' (e.g. Taylor and du Boulay, 1986; Fung, 1987).

The job now is to reconcile the factions. Foundations must be ascertained and verified in the context of utility. Models must be chosen that both afford the refinement of high-level reasoning and accommodate real-world constraints.

This chapter contrasts cursorily the aspirations of the general-purpose programming language designers with some evidence about expert problem solving and programming behaviour. The contrast is summarized in a rough 'wish list' of what experts want from general-purpose programming languages.

2 Design aspirations

Hoare wrote in 1973: '...good language design may be summarized in five catch phrases: simplicity, security, fast translation, efficient object code, and readability'. These provide a good centre for more recent aspirations, although, as Hoare anticipated: '...many language designers have adopted alternative principles which belittle the importance of some or all of these criteria, perhaps those which their own languages have failed to achieve'. Much language theory (e.g. denotational semantics) is concerned with separating the basic semantic description of a language from the notion, however abstract, of an evaluating mechanism or computer. This theoretical goal has infiltrated design attitudes: fast translation and efficient object code are 'implementation issues' considered separable from (and often subordinate to) qualities of the language model. Hence, these will not be discussed in this section. Hoare's three remaining 'catch phrases' will lead this (by no means exhaustive) discussion onto other design aspirations popular in the literature.

2.1 Simplicity

By simplicity, Hoare means a small range of instructions with uniform format, each having a simple effect that can be described and understood independently of other instructions. Terseness or simplicity of syntax is a common goal, one supported by studies of notations (e.g. Green, 1977) and reflected to some extent in recent designs (e.g. Miranda).

2.2 Orthogonality

Related to simplicity is orthogonality: the notion that there should not be more than one way of expressing any action in the language, and that all language components are mutually independent (cf. Denvir's more comprehensive examination of the notion, 1979). In a truly orthogonal language, a small set of basic facilities may be combined without arbitrary restrictions according to systematic rules. Orthogonality may result in arbitrary complexity via this unrestricted combination. It is endorsed in the literature in a limited role: as a principle for eliminating redundant expressions and hence for contributing to overall simplicity. This limited orthogonality is implied above in Hoare's meaning of simplicity.

2.3 Security

Hoare's principle is that only syntactically correct programs should be accepted by the compiler and that results (or error messages) should be predictable and comprehensible in terms of the source language program. No program should cause the computer to 'run wild' The classic examples for this aspiration are Pascal, a 'secure', strongly typed language, and C, a 'dangerous' one that produces occasional 'nasty surprises'. Building on the principle of security is correctness.

2.4 Correctness

This is the notion that a program can be proven to conform to a specification or to exhibit specified properties. This goal drives the aspiration to develop languages that conform to formal models and so are amenable to formal manipulation. Indeed, the pursuit of correctness is one of the *principal* aspirations of recent designs.

2.5 Readability

Hoare argues, simply, that programs are read by people, and that the programming language should encourage clarity. He asserts: 'The readability of programs is immeasurably more important than their writeability'. Unfortunately, a certain amount of myth adheres to readability: 'Authors choose stylistic factors commonly thought to influence style with little or no supporting evidence demonstrating the importance or effect of that characteristic on program comprehension and maintainability' (Oman and Cook, 1988). This notion leads to another: clarity of structure.

2.6 Clarity of Structure

Language designers aspire to express the intended solution structure visibly, without artificial imposition of extraneous structure, including artificial sequentiality. A related goal is to express any inherent parallelism (and to exploit it via the language implementation).

2.7 Modularity

One of the techniques for structuring larger programs is to segment or divide them into manageable pieces. Modularity allows programs to be segmented into independent chunks or *modules* which have *locality*, that is, which have internal scope and which interact with other modules via a defined interface. (cf., for example, Hughes, 1984.) The independence of the chunks is important, permitting separate compilation, separate debugging, independent testing, and reasonably easy replacement or modification. The independence and locality of modules is sometimes interpreted as 'hiding' an enforced rather than notional restriction of access to internals.

2.8 Abstraction

The motivation behind many language design decisions is to spare the programmer the 'messy' bits, the low-level, machine-oriented details of programming. The emphasis is on high-level operation, with the idea that high-level expression matches more closely the problem domain. As will be discussed later, this is a restrictive interpretation of abstraction.

3 The influence of the programming language on programming (that is, on devising solutions)

'Computer scientists have recognized that the features of a programming language can have a significant effect upon the ease with which reliable programs can be developed. It has also been observed that certain languages and language features are particularly well suited for the use of systematic programming techniques, while others hinder or discourage such discipline' (Wasserman, 1975, with reference to the then-recent structured programming techniques). The usual way to interpret this sort of statement is that the programming language guides the solution. Indeed, many of the arguments of modern designers are based on a conviction that programming languages influence the sorts of solutions that programmers will devise and that good languages guide the programmer's thinking (into whatever avenue is endorsed by the particular language model). Associated with this conviction is a certain amount of evangelism about how programmers *should* think.

Yet it is a vain hope that the programming language itself will provoke good code. (Just about everyone knows someone who can write Fortran in any language.) Flon's axiom (1975) states the case strongly: 'There does not now, nor will there ever, exist a programming language in which it is the least bit hard to write bad programs'. Experience with expert programmers (described in Section 4.2) offers an alternative interpretation of Wasserman's observations: that the programming language may facilitate clear *expression* of a solution, or it may obstruct it, but that the language does not influence strongly the nature of the solution strategy.

In the general case, language is a tool. Tools make easy the tasks for which they are designed, but the outcome depends on the intention and expertise of the wielder. Whereas a language can provide convenient categories, the support and heuristics for choosing the right abstractions in solving the problem are at best only implied. The *user* mediates the relationship between language and programming (i.e. devising solutions): individual expertise can compensate for deficiencies in language, whereas language cannot wholly compensate for deficiencies in use.

As summarized by Hoare (1981): 'I have regarded it as the highest goal of programming language design to *enable* good ideas to be elegantly *expressed*'. (Emphasis added.) Similarly, Dijkstra (1976) wrote: 'I view a programming language primarily as a *vehicle* for the *description* of (potentially highly sophisticated) abstract mechanisms'. (Again, emphasis added.) The goal of designers should not be to influence thinking but to reflect it clearly, to remove obstructions and restrictions. As Zemanek (1985) argues (as part of his case for formal definition): 'One cannot standardize thinking and one should not even attempt to do so'.

The following sections justify this more moderate view of the role of the programming language by considering some evidence about how experts solve problems and how expert programmers program.

4 How experts behave

4.1 Expert problem solving

Expertise in problem solving has been studied in a variety of domains, and results tend to be consistent across domains (for reviews, see Kaplan *et al.*, 1986; Allwood, 1986). Studies typically contrast novice with expert performance in solving problems. Experts differ from novices in both their breadth and their organization of knowledge: experts store information in larger chunks organized in terms of underlying abstractions. This organization apparently facilitates quick recognition of problem types and recall of associated solution strategies.

Experts sort problems in terms of underlying principles or abstract features (e.g. Chi *et al.*, 1981; Weiser and Shertz, 1983), whereas novices rely on surface features. Recall exercises (e.g. Chase and Simon, 1973; Shneiderman, 1976; Adelson, 1981, 1984; McKeithen *et al.*, 1981) have shown that experts are better able to reconstruct realistic domain configurations, such as chess positions or computer programs, although they perform no better than novices with random or scrambled stimuli. The interpretation for programming is that experts represent programs in terms of semantic structure, whereas novices encode them syntactically.

Experts tend to spend more time than novices planning and evaluating. Experts run programs more frequently while debugging than novices do (Gugerty and Olson, 1986). Experts are better able to form overviews, but thereafter they take longer to develop their understanding and representations, and they consider more fully interactions among functions or components of a system (Adelson *et al.*, 1984). Experts form a detailed conceptual model of a system and tend to incorporate abstract entities rather than the concrete objects specific to the problem statement (Larkin, 1983). Their models accommodate multiple levels and are rich enough to support mental simulations (Jeffries *et al.*, 1981; Adelson and Soloway, 1985; see also Chapter 3.1 concerning a schema-based view of expert programming knowledge and mental simulation).

The pattern overall is that experts are able to handle information at different levels. They differ from novices in two important respects: their ability to develop overviews or abstract models of solutions, and their ability to understand the consequences of implementation detail.

4.2 What experts do (and don't do) with programming languages

Given the emphasis, in the expertise literature, on abstract, semantic models rather than syntactic representations, where does the programming language impinge on programming?

One study of expert programmers solving problems in several general-purpose programming languages (Petre and Winder, 1988) has suggested that a programming language has only a weak influence on the solution. In this investigation, experts were presented with five non-trivial problems for solution in three languages each. The languages were chosen by the experts themselves, most choosing languages in more than one language style (e.g. Basic, Pascal, and Scheme; or Forth, KRC, and Prolog). In follow-up questionnaires and interviews, experts were asked to discuss, among other things, the suitability of each language for solving the problems or expressing the solutions.

Particular languages or language types did *not* correspond to particular solution strategies, although there *was* some correspondence between languages and oversights. It seems that experts conceive an abstract model of the solution separate from its expression in a particular programming language. There was evidence that they solve problems, not in the target programming language, but in a private, pseudo-language that is a collage of convenient notations from various disciplines, both formal and informal. The pseudo-language might be taken as the surface reflection of the expert's computational model, itself a composite of models borrowed from many sources.

Examples and remarks volunteered by the experts were sufficiently consistent to suggest the natures of these pseudo-languages. Variety of representation was important. Some of the notations derived from programming languages, including languages of different styles, others from mathematics, logic, natural language description, and sketches or diagrams – or combinations of these. Some code fragments were incorporated verbatim. The collections were apparently not made coherent but were assemblages of overlapping and possibly inconsistent fragments. The implication was that, although the pseudo-languages may not be self-consistent, coherence was imposed in use.

Different notations were used for different sorts of tasks. The associations between notations within the pseudo-language and general categories of problems and solutions were made on the basis of suitability; that is, the pseudo-language fragment made accessible some problem information or solution feature, or facilitated some set of operations, that was then of particular interest to the expert. A 'mix and match' description was typical, e.g. 'I find Algol-like pseudo-code useful for control-structure problems, together with arrays, sets, etc. from mathematics. This doesn't, however, cope well with control abstraction...for which I tend to use Scheme and Miranda-like notations...'. This sort of association is consistent with findings about expert problem classification (mentioned in Section 4.1) and makes good sense in terms of findings that problem representation is critical to ease of solution (See Section 2 on problem representation in Chapter 1.3). The borrowing in the pseudo-languages from many disciplines is suggestive; it may be that the pseudo-language provides a means of mapping or translation between problem and programming domains.

Given that the abstract algorithm is captured first in the programmer's own pseudo-language, expression in the target programming language is a matter of subsequent translation. This notion is by no means alien to the programming culture; Kernighan and Plauger (1974) endorsed it years ago as a principle for good design: 'Write first in an easy-to-understand pseudo-language; then translate into whatever language you have to use'. Hence, the algorithm is largely programming language independent, but it is dependent instead on some computational model private to the programmer.

This decoupling of programming – devising a solution – from programming language accounts for phenomena observed in the study. Experts were tenaciously resistant to a change of algorithm; in no case did a change of language *itself* provoke a change of algorithm. Unless provoked strongly, as by failure or inefficiency, experts did not look beyond the first algorithm 'good enough' to satisfy the problem. Rather, they were willing to tolerate heavy translation overheads in recoding a solution into an alternative language. Rather than guiding algorithm choice, programming languages may be distorted into conformance with the programmer's abstract solution.

In contrast, languages were changed much more readily than algorithms. Experts were willing to change languages just for fun as well as under the provocation of unacceptable obstruction (see next subsection).

4.3 Obstructions to coding; how programming languages get in the way

Apparently, although a programming language is unlikely to contribute directly to a solution, it may obstruct solution, even contributing to errors or oversights. Du Boulay and O'Shea (1981) also suggest that different error types may correlate to particular programming languages. In the study cited above (Petre and Winder, 1988), experts observed a number of coding obstructions.

Strongest among the experts' objections were obstructions to performance: inefficiency and an inability to access the facilities in the hardware. In other words, and in contrast to the aspirations of language designers, languages which did not support low-level manipulations were considered obstructive (e.g. 'I'd love to see someone try and do that in Scheme; you can't. You can't get at the bits.' and 'There's a lot of...design decisions which you can't easily express in a language which doesn't allow you to control storage'). All of the experts had an awareness of the machine that they instruct and, more broadly, of the world in which they operate. All of them claimed to have a reasoning model of the underlying machine; their aim was to produce a program that would run effectively on a computer. Many of the experts were not satisfied with a solution until the code had been compiled and run. Their concern about efficiency underlines the persistence with which reality constrains computing problems; there is a tension between the importance of the abstract algorithm and practical demands.

In order to satisfy their concerns about performance, experts demonstrated a wily evasion of restriction and a canny persistence in uncovering necessary details.

Other obstructions inhibited clarity of expression, a well-recognized goal: 'The problems that must be solved with today's languages are not simple. It is important that the programmer's task not be compounded by an additional layer of complexity from the very tool that is being used to solve the problem' (Marcotty and Ledgard, 1987). Experts bemoaned a lack of data structuring tools, the inadequacy of abstraction facilities, an inability to reflect the solution structure, and verbosity or notational clutter (e.g. 'So, how do you arrange, in a purely array world, to imitate a list?', '...you can actually manage to package up control structures in Scheme, whereas you can't really in C++ or Pascal. Now, the lack of that is something I've felt for a long time...' and '...the sheer textual complexity was enough to stop me even trying'). The conclusion was that 'ugliness' matters; programmers will refuse to use a programming language that is too obstructive or unpleasant. (Typical examples were Cobol, rejected for its verbosity, and Scheme, rejected for its messy punctuation.)

5 What expert language users want: a programming language 'wish list'

This section draws on the reported investigations of expert programmers to offer a wish list of general language facilities and qualities. Although this list is by no means comprehensive, it does attempt to capture the gist of what expert programmers expect from, and how they expect to use, general-purpose programming languages.

5.1 Granularity and control

The characteristic mastery by experts of both abstraction and detail, and the tension demonstrated by expert programmers between the abstract algorithm and practical demands, reflect the key tradeoff in modern programming languages: that between power, in the sense of abstraction and combination, and specific control. The experts want, on the one hand, the ability to abstract, to think in terms of high-level constructs, and, on the other hand, the ability to manipulate hardware or 'to get at the bits'. Experts need to look at different things at different times; in particular they need to choose their *grain of focus* at any one time (cf. Chapter 3.3 concerning opportunistic design strategies).

Expert programmers resent restrictions, particularly restrictions masquerading as 'protection'. The trouble in many languages is the assumption that programmers *can* forget details, that concrete detail, whether explicit or embedded in the language implementation, is 'accompaniment' to a program rather than part of it. These languages confuse 'exclusion of detail' with 'granularity of detail'. The notion of protecting the programmer from low-level details is, in the case of the expert, misguided. Lesk (1984) characterizes the problem: 'The good news about this is that you don't have to decide on the recursion or iteration yourself. The bad news is that you may not like what you get'.

A more appropriate aspiration than *protection* is *selection*: the ability to select the granularity of information for the context, when the context includes the intentions and goals of the expert as well as the nature of the problem, and to dip among the levels during a single task. The aspiration should be to provide information only when desired and to provide sensible defaults, so that, if the programmer is not interested, the implementation will handle the details.

Moreover, the language should provide comprehensive *control*. Experts want the language to be able to make full use of facilities in hardware, including interface facilities.

5.2 Basic, flexible tools

Expert programmers are pragmatists. They tend to rely on the language features they can count on getting. Like smiths, they expect to create most of their own tools. Discovering and remodelling hidden, high-level mechanisms is perceived to entail more work and esoteric knowledge than building utilities up from a low level. Hence, experts tend to prefer generality and flexibility to specific in-built power, although they will make use of appropriate tools when convenient.

Much of the enthusiasm for the newer programming styles (e.g. functional, object oriented) is accounted for as recognition rather than conversion. Programmers receive gladly tools for doing what they intend and have been achieving by less direct routes, i.e. by writing programs in a style other than that of the programming language (e.g. 'structured' assembler, or 'functional' C), or by implementing the desired evaluation mechanisms in whatever language is available. One common pattern is to use some crude but effective language (e.g. C) to build an intermediate language customized for the problem domain (e.g. writing an object-oriented superstructure for C, or a schematics-encoding language in Lisp).

But experts don't view handy facilities as substitutes for basic utility; these same celebrants argue against restriction. Often they achieve the desired combination of expressive power and control by mixing languages (e.g. writing the bulk of a prototype in Prolog, but handling the input/output and graphics in C).

Implicit in arguments for granularity of focus and for flexibility is recognition both that any general purpose language can express any solution expressible in another, and that each language expresses some things well at the expense of others.

5.3 Structural visibility, modularity, and abstraction

The desire for abstraction facilities extends in part from the desire for structural visibility. Experts want to be able to express the solution structure as they perceive it, with support for the conception and amendment of that structure. They want structuring tools both for expressing strategy and for representing data. They want tools for packaging mechanisms into structurally and notionally sensible parcels (or 'modules' with locality (that is, with internal scope and a defined interface to other parcels). Further, they want tools for abstraction, so that they can generalize parcels into program concepts to be used and reasoned about in a way that de-emphasizes their internals.

Reiterated within this sense of abstraction is the argument about granularity of focus. A choice of granularity leaves the constructs within a program abstraction beneath the level of concern. It is critical to distinguish the choosing of focus from invisibility. The distinction is between the option to ignore the detail encompassed within an abstraction, and inaccessibility; i.e. between 'needn't look' and 'cannot see'.

Such powerful structuring tools are not without their dangers. Abstraction building is seductive; forming generic abstract types can lead into confusing excess and even eclipse the problem itself. It is not clear that the languages which offer abstraction tools actually help in choosing – and choosing the extent of – abstractions, or facilitate rebuilding if the abstraction structure fails.

5.4 Accessibility

Experts want access to a complete, workable model of the language, preferably without digging into the implementation. In general, computational knowledge, not machine knowledge, is critical to solution (e.g. an effective knowledge of what a stack is and how it operates need not include the details of the machine instructions and memory locations used to implement it). Unfortunately, computational knowledge is still often unavailable except via hardware or language implementation knowledge. The notion of 'freeing the programmer' from implementation details is misguided when, as in the case of hidden evaluation mechanisms, the language model is not deducible from the language surface, or when, by reading a program, the programmer cannot anticipate what the machine will do.

In the case of the more 'powerful' declarative or specialist languages (e.g. Prolog, SQL), some algorithmic decisions are pre-empted by the language implementation. That is, the language implementation injects algorithmic information not in the program, so that part of the solution is embodied in the language. To that extent, the language contributes directly to the solution, and the programmer requires at least an abstract knowledge of those in-built mechanisms in order to comprehend a solution strategy and its behaviour completely.

In order to code effectively in a language, the programmer needs a fairly accurate model of the language for mapping with the solution model. The idea of separating the language definition and design from the language implementation is a good one – but if a complete, workable model of the language is not accessible via an alternative source, the programmer will delve through the implementation and will exploit whatever 'hacks' were needed to make the language work.

5.5 Predictability and efficiency

Experts need to be able to predict, at least roughly, the behaviour of their programs, including efficiency, side-effects, by-products, and data end-state. Wirth (1974) called this *transparence*, and considered it an aspect of simplicity: 'Do not equate simplicity with lack of structure or limitless generality, but rather with transparence, clarity of purpose, and integrity of concepts'. By transparence, Wirth means that a language feature 'does not imply any unexpected, hidden inefficiencies of implementation' and that 'the basic method of implementation of each feature can be explained independently from all other features in a manner sufficiently precise to give the programmer a good estimate of the computational effort involved'.

Efficiency, far from being an 'implementation detail' separable from program – and language – design, is of primary importance to expert programmers. 'What seems clear is that people settle on fairly efficient languages, regardless of the claims of the high level language designers' (Lesk, 1984).

5.6 Conciseness or simplicity

Experts appreciate an uncluttered notation. As reported earlier, 'ugliness' matters.

6 Summary and conclusion

The list of what expert programmers want in a general-purpose programming language differs from the aspirations of language designers less in detail than in emphasis. The well-foundedness versus utility schism is reflected in the designers' emphasis on correctness and the expert programmers' emphasis on predictability and efficiency. To exaggerate the point: whereas designers often aspire to provide a panacea, users want tools. Although programmers do not object to the advantages of formal manipulation, they will not, in general, accept them at the expense of control and performance. A second important discrepancy is in the interpretations of abstraction: what designers discuss in terms of hiding and restriction, programmers desire as a selection of focus paired with sensible defaults.

What expert programmers seem to want is what Hoare prescribed back in 1973: 'A good programming language should give assistance in expressing not only how the program is to run, but what it is intended to accomplish; and it should enable this to be expressed at various levels, from the overall strategy to the details of coding and data representation'. Or, as Tennant (1981) phrased it: 'In short, an ideal programming language would combine the advantages of machine languages and mathematical notations...'.

Programmers often address inherently complex tasks. If the programming language is itself complex, it may be an obstruction rather than a tool; it may contribute to the problem instead of to the solution. The point remains to provide good tools for people who instruct computers, that is, tools for tool-users.

References

Adelson, B. (1981). Problem solving and the development of abstract categories in programming languages. *Memory and Cognition*, **9**, 422-433.

Adelson, B. (1984). When novices surpass experts: the difficulty of a task may increase with expertise. *Journal of Experimental Psychology: Learning, Memory, and Cognition*, **10**, 483-495.

Adelson, B. and Soloway, E. (1985). The role of domain experience in software design. *IEEE Transactions on Software Engineering*, **11(11)**, 1351-1360.

Adelson, B., Littman, D., Ehrlich, K., Black, J. and Soloway, E. (1984). Novice-expert differences in software design. *Interact '84, First IFIP Conference on Human-Computer Interaction.* Amsterdam: Elsevier.

Allwood, C.M. (1986). Novices on the computer: a review of the literature. *International Journal of Man-Machine Studies*, **25**, 633-658.

Chase, W.G. and Simon, H.A. (1973). Perception in chess. *Cognitive Psychology,* **4,** 55-81.

Chi, M.T.H., Feltovich, P.J. and Glaser, R. (1981). Categorization and representation of physics problems by experts and novices. *Cognitive Science,* **5,** 121-152.

Denvir, B.T. (1979). On orthogonality in programming languages. *SIGPLAN Notices,* **14(7),** 18-30.

Dijkstra, E.W. (1976). *A Discipline of Programming.* Englewood Cliffs, NJ: Prentice Hall.

du Boulay, B. and O'Shea, T. (1981). Teaching novices programming. *In* M.J. Coombs and J.L. Alty (Eds), *Computing Skills and the User Interface.* New York: Academic Press.

Flon, L. (1975). On research in structured programming. *SIGPLAN Notices,* **October 1975,** pp. 16-17.

Fung, P. (1987). Novice PROLOG programmers: a consideration of their problems. CITE Technical Report No. 26. Milton Keynes: Open University.

Green, T.R.G. (1977). Conditional program statements and their comprehensibility to professional programmers. *Journal of Occupational Psychology,* **50,** 93-109.

Gugerty, L. and Olson, G.M. (1986). Debugging by skilled and novice programmers. *Proceedings CHI'86: Human Factors in Computing Systems.* New York: ACM.

Hoare, C.A.R. (1973). *Hints on programming language design. SIGACT/SIGPLAN Symposium on Principles of Programming Languages,* October 1973.

Hoare, C.A.R. (1981). The emperor's old clothes. *Communications of the ACM,* **24(2),** 75-83.

Jeffries, R., Turner, A.A., Polson, P.G. and Atwood, M.E. (1981). The processes involved in designing software. *In* J.R. Anderson (Ed.), *Cognitive Skills and Their Acquisition.* Hillsdale, NJ: Erlbaum, pp. 255-283.

Kaplan, S., Gruppen, L., Leventhal, L.M. and Board, F. (1986). *The Components of Expertise: A Cross-Disciplinary Review.* Ann Arbor, MI: The University of Michigan.

Kernighan, B.W. and Plauger, P.J. (1974). *The Elements of Programming Style.* London: McGraw-Hill.

Kowalski, R. (1982). Logic as a computer language. *In* K.L. Clark and S.-A. Tarnlund (Eds), *Logic Programming.* London: Academic Press.

Larkin, J.H. (1983). The role of problem representation in physics. *In* D. Gentner and A.L. Stevens (Eds), *Mental Models.* Hillsdale, NJ: Erlbaum.

Lesk, M. (1984). Programming languages for text and knowledge processing. *In* M. Williams (Ed.), *Annual Review of Information Science and Technology,* vol. 19. Knowledge Industry Publications, pp. 97-128.

McKeithen, K.B., Reitman, J.S., Reuter, H.H. and Hurtle, S.C. (1981). Knowledge organisation and skill differences in computer programmers. *Cognitive Psychology,* **13,** 307-325.

Oman, P.W. and Cook, C.R. (1988). A paradigm for programming style research. *SIGPLAN Notices,* **23(12)**, 69-78.

Petre, M. and Winder, R.L. (1988). Issues governing the suitability of programming languages for programming tasks. *People and Computers IV: Proceedings of HCI'88.* Cambridge: Cambridge University Press.

Shneiderman, B. (1976). Exploratory experiments in programmer behavior. *International Journal of Computer and Information Sciences,* **5**, 123-143.

Taylor, J. and du Boulay, B. (1986). Why novices may find learning PROLOG hard. *In* J. Rutkowska and C. Crook (Eds), *The Child and the Computer: Issues for Developmental Psychology.* Chichester: Wiley.

Tennant, R.D. (1981). *Principles of Programming Languages.* Englewood Cliffs, NJ: Prentice-Hall.

Wasserman, A.I. (1975). Issues in programming language design–an overview. *SIGPLAN Notices,* July 1975, pp. 10-12.

Weiser, M. and Shertz, J. (1983). Programming problem representation in novice and expert programmers. *International Journal of Man-Machine Studies,* **19**, 391-398.

Wirth, N. (1974). On the design of programming languages. *Proceedings of IFIP Congress 74.* Amsterdam: North-Holland, pp. 386-393.

Zemanek, H. (1985). Formal definition the hard way. *In* E.J. Neuhold and G. Chroust (Eds), *Formal Models in Programming.* Amsterdam: Elsevier, pp. 411-417.

Chapter 2.2

Programming Languages as Information Structures

T.R.G. Green

MRC Applied Psychology Unit, 15 Chaucer Road, Cambridge CB2 2EF, UK

Abstract

This chapter describes three 'implicit theories' of programming, which have at different times governed the styles of programming language design. Not surprisingly, the experiments performed by empirical researchers have also been deeply influenced by the prevailing theory of the day. (1) The first view regarded programming as transcription from an internally-held representation of the program. This leads to Fortran-like languages, and the early empirical studies of programming correspondingly emphasised particular language features and their incidence of errors. (2) The second view stressed program comprehension and required programs to be demonstrably correct, leading to Pascal-like languages and the pseudo-psychological theory of 'structured programming'. Empirical research during this era showed that giving programmers easy access to information they needed (whether structured or not) was what really mattered. Visual programming began during this era, but most empirical studies failed to demonstrate any inherent advantage to visual presentation. Where advantages were found, they appeared to fit the same fundamental principle. (3) Today's view is that program design is exploratory, and that designs are created opportunistically and incrementally. Supporting opportunistic design means putting a minimum of unnecessary demands on working memory, permitting designers and

Psychology of Programming
ISBN 0-12-350772-3

programmers to postpone decisions until they are ready for them, allowing easy additions or changes to existing code, etc. Empirical research on change processes and reuse of code has demonstrated differences between language designs, but has a long way to go. The chapter ends with suggestions for 'lowering the cognitive barriers to programming' by more careful design of languages.

1 Introduction

Even a decade ago, a survey distinguished about 150 documentation techniques (Jones, 1979). There are probably far more now – and far more programming languages. They vary a great deal in every possible respect. Even for experts, certain details of language design – what I shall call the 'information structure' – are likely to affect the speed and accuracy of using the language or the documentation technique. Any readers who belittle the relevance of notational structure, and prefer to believe that 'anyone can get used to anything', are recommended to earn their living doing arithmetic with roman numbers for a while. So, whereas other chapters (e.g. Chapters 1.3, 2.4 and 3.2) describe what we know about the *mental processes* of programming, this chapter describes research on how *external factors* influence the processes.

1.1 Information structure

We can consider the user of a programming language as someone who has to find out information from a program, add new information to it, and occasionally reorganize it; and the details of the programming language will influence how easy it is to do such tasks. The framework we shall adopt is that a program is a kind of information structure, just as a library is a kind of information structure. Libraries can be organized in many ways, and different choices will facilitate different tasks. Most libraries determine the shelf positions of books exclusively by subject indexing, but a few use a different principle: in Cambridge University Library, for instance, shelf positions are determined partly by subject matter but also by the physical size of the book (so books of different size categories are stored separately). Both these systems can be used to shelve the same books – in other words, they have equivalent power; what is different is the way the information about the books is structured. In consequence, different sets of user tasks are supported by the systems. Browsing by subject in Cambridge University Library is notoriously unrewarding – but *if* you happened to want unrelated books of the same size, it might be just what you wanted.

One of the themes of the chapter is that an information structure can make some information more accessible, but usually at the cost of making other information less accessible. Programs have in common with libraries and other information structures that *the structure of the information should match the structure of the task*.

Given this outlook, I shall not make much distinction between a true programming language, on the one hand, and a documentation format, on the other. The principles governing access to information, ease of change, etc., will presumably be the same.

In subsequent sections we shall see how the view of the programmer's task has altered over the years. The design of programming languages has altered in sympathy with different views of the task, and the style of programming research has likewise altered.

1.2 The state of evidence

The aim of this chapter is to bring together the existing empirical evidence, rather than to present responses from practitioners, however revealing. Much of the available evidence was gathered in pursuit of some other objective than comparisons of notational structure, especially in studies of the problems of novice programmers. It is still relevant but needs caution: learners and experts are likely to have different notational needs. One would expect that while the problems of learners are often to do with recognizing and recalling components, with fitting them together and with perceiving relationships, the problems of experienced users would be 'high-performance' problems – finding information quickly, and modifying programs without unnecessary effort. (Comparisons of student and professional performance by Holt *et al.*, 1987, support this view.)

Rather than painfully distinguish all through the chapter between the problems likely to affect learners and the problems likely to affect experts, I have simply tried to clarify concepts and sources of difficulty. Anyone evaluating a programming language for actual use will have to consider many criteria; among them should be sources of difficulty *for the intended users*.

2 The programmer's task

2.1 Types of task

Consider, therefore, what tasks are required of programmers. Present-day views have been heavily influenced by the work of Sumiga and Siddiqi, Visser, and Guindon (see Chapter 3.3) showing clearly that the design of large programs by professionals corresponds to 'opportunistic planning', an activity in which a design is constructed piecemeal with frequent redesign episodes. Much use is made of an external record, whether on paper or a VDU, and new portions are inserted as they come to mind. As far as we can tell at present, this picture applies in any activity which has a large design component, regardless of the notation or of the stage in the development process.

Now, this is *not* the 'approved' style of development. Software designers like to speak of the 'waterfall' model, in which higher-level requirements are dealt with before starting on lower-level processes; or of other similar methodologies. Perhaps such disciplined approaches would be better, but they seem to be infeasible as a general technique, because the consequences of higher-level decisions cannot always be worked out fully until lower level ones are developed. Be that as it may, this chapter will concentrate on what it seems that people really do, not on what they might do if they were perfect.

Typical activities therefore include: comprehending or 'parsing' the partially developed design, to remind oneself what has been written so far and how it works; making modifications, of any scale, large or small; inserting new components into what exists already; looking ahead to foresee consequences of design decisions (this is often done in an ancillary formalism of some sort – see, for example, self-observational studies by Naur, 1983); and recording degree of commitment to particular design choices.

Although typical experimental studies do not, unfortunately, record such a wide variety of activities, the importance of discriminating between them has been demonstrated, at least for the comprehension case. Comprehension has been shown to be a complex task (see Chapter 1.3), of which one facet is conveniently labelled 'deprogramming', meaning that after a portion of the mental representation of the problem has been translated into code (or specifications language, or some other notation), it is then translated back again into mental representation language, as a check. In many notations there is an apparent asymmetry of effort: it seems to be easier to develop the code than to recover its meaning. (Spreadsheets, for example, are created quickly and easily, but discovering precisely what a spreadsheet does can be difficult.) In other notations, the asymmetry is much less, and 'deprogramming' is relatively straightforward.

To appreciate how deprogramming was identified as a problem we need to look ahead a bit and mention some studies; we shall return to them in their proper place. The asymmetry of effort between writing and comprehending was first postulated by Sime *et al.* (1977), after they had shown that novices debugged their own programs much faster in one miniature language than in two others. The languages were based on conditional structures, and it was claimed that the deprogramming problem lay in working 'backwards' through the program, to discover what set of circumstances ('taxon') caused a particular outcome. Later, Green (1977) strengthened that conjecture, by showing that professional programmers could trace one type of conditional program backwards much faster than another, even when the forwards tracing times were not significantly different. The clincher was given by Curtis *et al.* (1988), who showed a statistical separation between different types of question in the same experiment, and thereby demonstrated that the concept had discriminant validity. Their study also showed that understanding data flow was a third type of task, again statistically separable from the previous two.

What all this shows, therefore, is that detailed aspects of information structure can have large effects on particular parts of the process of creating designs and programs. Other parts of the process may be almost entirely unaffected. To discover the virtues or difficulties inherent in particular notational structures it is therefore necessary to consider many different tasks.

2.2 How information structure affects behaviour

Differences in information structure are sure to affect the programmer, but not necessarily directly. Green (1989) has used the idea of opportunistic planning to develop an analysis of 'cognitive dimensions' of notations, with suggestions as to how variations along each dimension affect typical programming tasks. (More accurately, what affects the usability is the combination of a notation and the environment in which it is used: a notation suitable for 'pen and paper' might be less suitable for a top-down structure-based editor.) We cannot investigate these dimensions in any detail here, but we can give a few examples.

The first is *viscosity*, or the degree of resistance to local changes. The degree of viscosity clearly depends on what change is being made, for we can see that it is often very easy to make a minor change to a relatively loose structure, like Basic, but a

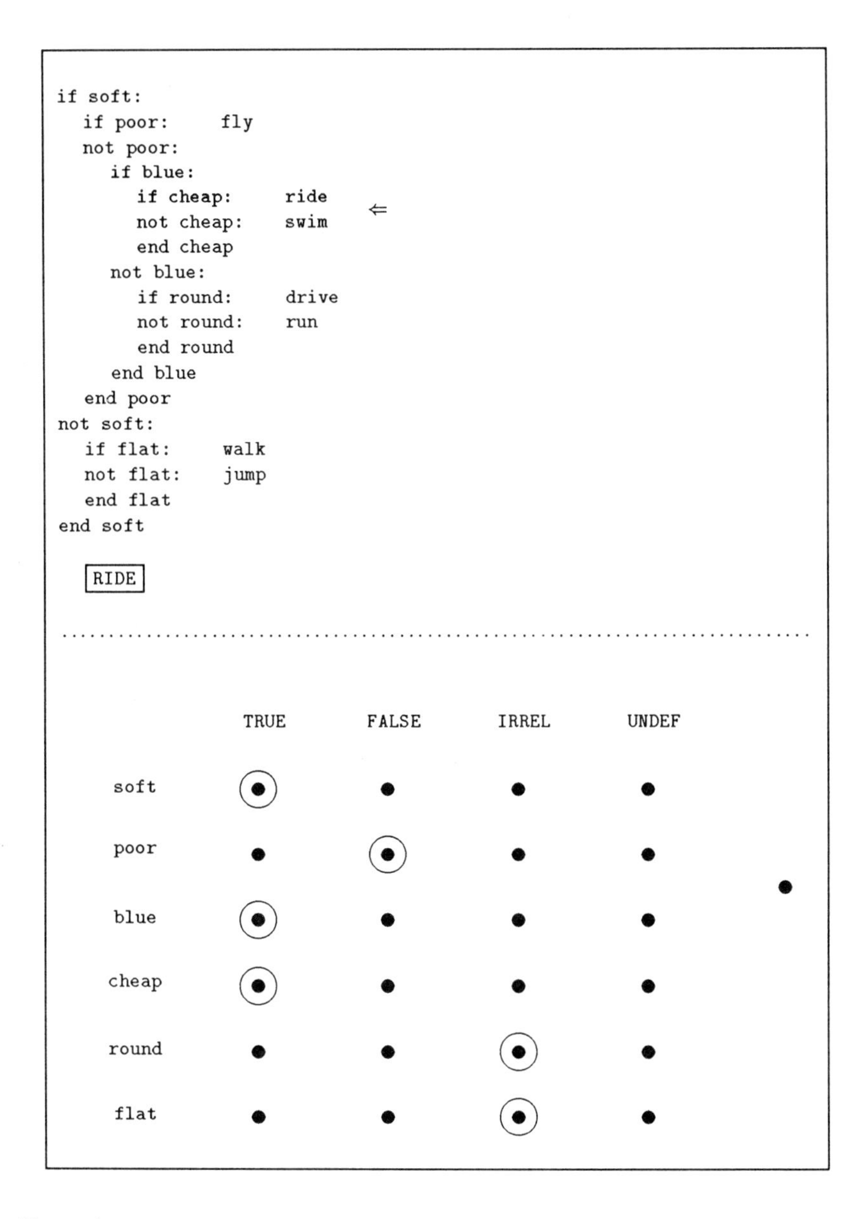

Figure 1: An experimental task from Green (1977), simulating the postulated mental activity of 'deprogramming'. Given the program (*upper panel*), the subject's task was to report the conditions under which the designated action 'ride' will be the first action performed. Responses were made using a stylus, touching studs on a response display (*lower panel*). Here, the full response has been set up; touching the far right-hand spot will complete it. Subjects who were professional programmers found this task much easier in this type of miniature language than in the types based either on GOTOs or on conventional nested structures.

large change tends to create 'knock-on' effects, further changes that are consequences of the original one. The object-oriented programming systems are particularly effective in reducing knock-on effects, and to that degree are an improvement on the dimension of viscosity. A second example is *premature commitment*, where the programmer is forced to make a decision before the consequences can be foreseen. This can happen where the notation and the editor, or some other construction system, are incompatible, or where the system has been designed around a different view of programmers' tasks. Hoc (1981, 1988) has demonstrated the problems that arise with editing systems that enforce commitment to outline program designs before the programmer is fully ready. It is noteworthy that many designs for structure-based editors, despite being intended to help programmers, seem likely to demand premature commitments. The third example is *role expressiveness*, the ease of 'de-programming', discovering what the parts of an existing program are and what is the role or purpose of each part. An excellent study by Pennington (1987), described by Gilmore (in Chapter 1.5), has demonstrated notational differences in role expressiveness, by showing that role-based abstractions are more easily constructed about Fortran programs than about Cobol ones.

The suggestion is that programmers will choose their style of working according to the particular combination of information structure and editing tools. If the system is viscous, they will attempt to avoid local changes and will therefore avoid exploratory programming. If the system has poor role expressiveness, at least the less experienced programmers will avoid exploratory programming, since they will have trouble in recognizing program components. If the system enforces premature commitment, the effect will depend on the viscosity; with low viscosity, premature commitment is only a small risk; as viscosity rises, postponing commitment becomes more urgent.

2.3 Implicit theories of low-level programming

Clearly, if we take opportunistic planning seriously, the demands on the notation are subtle and various. Both language designers and language researchers have taken a long time to reach today's views, and no doubt further complexities are still to come! But to understand the research tradition it is necessary to realize that the generally accepted view of programming has changed over the years, and with it the focus of empirical research has also shifted. Language designs attempt to meet the designer's view of programming, and so the design presents us with an 'implicit theory' of what tasks the programmer must accomplish.

Three main stages can be distinguished in the implicit theories apparent in programming languages. First, the implicit theory was that programming was an errorless transcription. So long as a program performed the correct computation, its comprehensibility was immaterial; no thought at all seems to have been given to problems of the modifying programs. In the second stage, language designers gave more attention to programs as constructions that had to be comprehended by others, or possibly by programmers themselves at some later date. Languages conforming to this view still paid little attention to the modification of programs; they were presented as solutions to well-defined problems, and much was said about the desirability of knowing that the solution was correct before starting to code the program. In the third stage, where we are at present, many designers accept the 'evolutionary' style of programming, in which the activity of program design is one of repeated modification – frequently starting from a seed which was an already-existing program that solved a related problem.

Since the prevailing view has affected the type of research question that empiricists asked, we shall review the research under appropriate headings.

3 The errorless transcription view of programming

3.1 The implicit theory

We will take the Fortran-Basic tradition as a model of how programming languages were originally designed. The implicit theory could be described as the 'one-way, error-free' view. Fortran I was a very great success in its day, showing that the implicit theory was clearly at least partially correct – in its day.

The one-way view of programming is very simple. It sees programming as *errorless transcription from a previously worked-out representation* (possibly held in the head, possibly on paper). The mental representation of a program is apparently viewed as a sequence of steps – step 1, step 2, step 3, ... – and each step is individually translated into its coded representation in the target programming language. Nothing more is involved. Consider some of the design features of this tradition:

* The Fortran programming system (punched cards) and the Basic line-numbering system encouraged programmers to create their programs in the order of the text – i.e. line 1 of the final text was also the first line to be punched in. Thus, *the program was fully developed at the start of coding, needing only to be transcribed.*

* Fortran and Basic have very few guards against typing errors, which can readily create a new text that is syntactically acceptable but not, of course, the intended program. By implication, *programmers do not make typing errors.*

* Neither Fortran nor Basic originally supported any use of perceptual cues to help indicate structure. Possible cues would have included indented FOR-loops, demarcated subroutines, bold face or capitals to indicate particular lexical classes, etc. The implication is that *programmers can comprehend the program text without assistance.*

* The use of GOTOs as the sole method to determine control flow encourages small changes but makes large changes extremely tedious. The implication here is *programmers do not need to modify their first version, except trivially.*

Fine, you think. All that used to be true, but we have moved on. But have we? Consider the structure of spreadsheets (a form of programming language), of Prolog, and of the production system languages used in expert systems. In many important respects these three systems continue the tradition of the 'one-way' implicit theory. None of them guard against typing errors, none of them supply perceptual cues, none of them limit the complexity of program structures that have later to be understood. Prolog and production systems allow the text to be generated in the order the programmer prefers, rather than starting at line 1, but spreadsheets encourage starting at the top left-hand corner, whether convenient or not. One advance that has been made is that subsequent modifications are sometimes easier. Not always: some types of modification are really long winded, in each of these three languages. Many other systems could be pointed to. In short, the implicit theory of programming as transcription is still alive.

3.2 Research on language features

Empirical research arising from this view of programming is likely to focus on the individual features or syntax constructions of programming languages. From the 1970s onwards there have been several such studies. Youngs (1974) reported frequencies of errors for a number of different statement types (comments, assignments, iterations, GOTOs, conditionals, etc.) in several languages. These results were achieved simply by looking at programs that had been written and looking for where the bugs occurred. Although his study was more comprehensive than most similar studies, it is difficult to see what has been learnt from a collection of error frequencies that can usefully be generalized to the problem of language design. Slightly more can be learnt by comparing different designs for the same language feature, such as logical versus arithmetic IF statements in Fortran (Shneiderman, 1976) or nesting versus GOTO styles of conditional (Sime *et al.*, 1973, 1977; Mayer, 1976).

Many of the early studies asked simplistic questions: 'Are logical conditionals [always] better than arithmetic ones?', 'Are flowcharts [always] better than code?', 'Are nested conditionals better than GOTOs?', etc. It is easy to see today that in general the answer is going to be 'X is better than Y for some things, and worse for others'. But that is because we can clearly see now that programming is a complex set of skills, not a unitary process.

In so far as a unitary process can be expected, it might be 'readiness to build a program rather than use repeated operations'. In a thorough set of investigations, Wandke and his colleagues have investigated changes in programming readiness during the acquisition of expertise in very simple situations. In much of their work, subjects practised searching a database for various targets, using a miniature search language. Typically, the language contained five commands plus one command to define a macro-instruction. A carefully designed study (Wandke, 1988) showed that subjects were reluctant to define macros for conditional constructions even when very large numbers of keystrokes in subsequent search operations would have been saved. The inclusion of control structure dramatically increased cognitive effort; simply counting keystrokes gives no indication of the real effort involved.

In a further study from the same group, Wetzenstein-Ollenschlaeger and Schult (1988) also put subjects in a situation where it would save effort to build macros, and after each of four sessions tested subjects' knowledge of semantics and syntax and of the interrelationships between the commands in the miniature language. The task structure had a core command sequence, which had to be repeatedly typed out in full unless a procedure was defined for it. Numbers of subjects using procedures rose with experience. Subjects using procedures scored slightly higher on knowledge of syntax and semantics, though not much. But they scored a great deal higher on knowledge of interrelationships between commands (about 70% against 11 to 50%). The interesting fact is that the bulk of this difference came not from the core task, but from the other parts of the task – the commands that were *not* proceduralized. Readiness to use procedures, in short, depended on understanding the whole of the language, not just the immediately relevant parts.

4 The 'demonstrable correctness' view of programming claims

4.1 The implicit theory

With the rise of 'structured programming' in the 1970s a new implicit theory emerged: programs had to be clearly seen to be correct. Comprehensibility was equated with formal structural simplicity, which favoured hierarchical composition, and harsh words were used of earlier programming languages ('Basic causes brain damage', 'Fortran programmers can never learn sound programming principles'). Programs were to be built from a small number of structures which could be related to each other in simple ways. This was the hey-day of the 'neats' (see Chapter 1.2).

Pascal, a powerful influence, rejected GOTOs in favour of a small repertoire of nestable loop and conditional structures, and also contributed a technique for defining hierarchically composed data structures. After Pascal, further developments down the 'neat' line rejected variables and iterative constructions in favour of pure compositions of functions. The argument in all cases was that the correctness of programs was easier to perceive.

4.2 Research on program structures and the structured programming claims

The second phase of empirical research on programming can be seen as confronting the structured programming dogma with empirical data which showed that the story was more complex. One line of research concentrated on the recommended style of program construction. The structured-programming methodology of 'stepwise refinement' instructed the programmer to decompose difficult problems into easier problems, and then into still-easier problems. For this reason, hierarchically structured notations were preferred, and much opprobrium was poured on users of notations that allowed 'unstructured' control flow. Doubts were soon raised as to whether research on problem solving really supported the enforced use of hierarchical structures – among other questioners, Green (1980) contrasted the computer scientists' assertions with evidence that showed that problem solvers did *not* customarily use such simple methods as stepwise refinement.

These doubts were matched by empirical evidence. Sime *et al.* (1977) compared novice programming in three types of notation, one of which was unstructured, one of which was structured in the usual way, and one of which was structured in the same way but also contained additional information. The additional information was logically redundant, since it could be deduced from what was already present. They showed that both the nested notations were improvements over the unstructured notation, as predicted by the structured programming school, but that the additional information greatly helped novices to find solutions to short programming problems using conditional structures. In particular, as mentioned above, it greatly aided 'de-programming'. Thus the structured programming approach had grasped only some of the truth. Another study (Arblaster *et al.*, 1979) explored the difference between structured and unstructured notations. Was it vital to have a hierarchical structure, as claimed by structured programmers, or would other types of structure also be effective? Their results showed that of three types of structuring (hierarchical, decision-table-like, and compromise), all were better than a condition with no structure but that the hierarchical structure was not markedly better than the other two.

Once again it appeared that the structured programming school had only grasped some of the truth.

An important qualification was made by Vessey and Weber (1984), showing that it was possible to differentiate between effects due to problem solving and effects due to coding. They held the problem-solving component constant by presenting problems in a neutral language, and extended Sime and co-workers' three languages to include unindented and indented forms. (Sime *et al.* had studied indented nested languages versus unindented GOTO languages.) Apparently novices' performance was determined much more by indentation than Sime *et al.* had supposed, and the relative advantages of nested conditionals were less clear cut.

Laboratory-based research using 'micro-languages' may not readily generalize, of course, but at least it was shown that these results generalized to professional programmers: Green (1977) showed that the hierarchical structure with added information allowed professional programmers a small but highly significant speed advantage in answering certain types of questions about programming. This result shows that the 'coding' explanation advanced by Vessey and Weber is not entirely sufficient.

A more important aspect of Green's study was that the task of 'program comprehension', which had previously been treated rather casually, was given a deeper analysis. Many previous studies had asked subjects to execute the program mentally and to report on what output was achieved for given input. Green added the inverse, or 'deprogramming', task of asking what input was required to achieve given output (see Figure 1). It was this second task that differentiated between the different notational structures. The results obtained earlier by Sime *et al.* (1977), which had shown that novice programmers debugged their programs faster in the language with additional information, were therefore ascribed to differences in the ease of deprogramming the different notations.

4.3 Generalizing to other computational models

The explanations advanced by most researchers studying notational design have been phrased in terms of information-processing demands. In principle, therefore, they could be applied to any programming paradigm, not only to procedural programming languages. Gilmore and Green (1984) set out to demonstrate a parallel of Green's procedural-language results, this time using a declarative programming system. They based their study on what looked like a very obvious claim: if it was easier to answer procedural-type questions than declarative ones, given a procedural program, then it should be easier to answer declarative-type questions than procedural ones, given a declarative program.

Looking at their study in more detail, Gilmore and Green argued that Green's 'deprogramming' task was equivalent to asking subjects to convert from a procedural representation into a declarative representation. The additional information given in one of Green's notations assisted the subjects because it contained cues to the declarative form. The other two notations contained no such cues and hence answering the declarative-type questions was harder. *Mutatis mutandis*, similar results should be obtained by starting with a declarative notation: procedural questions would be harder than declarative ones, since they would require information in a structure not provided by the notation. The extra difficulty would be lessened if additional information were given, containing cues to the procedural form. So they compared the difficulty of answering both declarative questions and procedural questions, working from declarative programs with and without cues to procedural information.

The results were, however, less clear cut than was expected: the anticipated effects were present but weak, except when the subjects were working from memory, in which case quite effective results were obtained. Why should this be? We do not, at present, know: but at least two factors could contribute. The first is that the 'natural' mental representation for the type of problem used may often be procedural rather than declarative. This has been suggested by work such as that of Hoc (see Chapter 2.3) which indicate that novices frequently conceive programs as a series of steps rather than as what-if structures. The second factor may be simply that the cues used in Green's study were more perceptual in nature, while the Gilmore and Green study used cues that were more symbolic in nature. The importance of perceptual cues to program structure will be taken up below.

4.4 Program visualization: diagrammatic notations

Many attempts have been made to replace text-based programming notations by diagrammatic notations, in the hope of improving comprehension. Reviewing them will take a rather long section because a subsidiary argument must be included: whether one must use genuine diagrams, proper little pictures; or whether perceptual enhancements to program text will be adequate. Is there evidence that diagrammatic notations genuinely offer something that symbolic notations cannot match? If so, what; and how do we balance the trade-off against the disadvantages of having no text editor, no easy electronic mail, and all the other infra-structure of text? This is a crucial issue, at present little understood. It is disturbing to observe that many workers have not even asked the question, and instead assert uncritically that 'Pictures are more powerful than words ... Pictures aid understanding and remembering ... Pictures do not have language barriers. When properly designed, they are understood by people regardless of what language they speak' (Shu, 1988, pp. 7-9).

If the more dewy-eyed propositions of program visualization enthusiasts are accepted, then we can expect diagrams to be better for all purposes (and enhanced textual presentations to offer no advantages). The alternative position, more in keeping with the outlook of this chapter, is that the critical factor determining comprehensibility of notations is *accessibility of information*. Certain types of information are likely to be accessed more easily from diagrams. In many cases, however, perceptual enhancements to text-based presentations will also improve access.

(We should also note that all the questions that might affect choices in a real world have been brusquely suppressed. What about the sheer physical size of a diagrammatic representation of a complex program, or the memory requirements, the editing facilities, the speed of display, compilation and execution? We leave all those aside so that we can concentrate on the prior question: are diagrams any good?)

Although developments in program visualization are a lively area, and have given rise to various reviews and taxonomies (Myers, 1986; Shu, 1988; Chang, 1990) and some special issues of appropriate journals, we have at present nothing remotely resembling an adequate body of empirical research. There is clear evidence that both diagrams and enhanced text presentations can improve performance, but we do not at present know the types of task that are most improved. We shall consider diagrammatic notations in this section, and enhanced symbolic notations in the following one.

Despite the wide variety of diagrammatic representations that have been proposed, almost all the existing empirical studies deal with comprehensibility of flowcharts, a notation now widely believed to be rather poor. Early studies tended to support that opinion. Thus Shneiderman *et al.* (1977) found little advantage for novices in having a flowchart as supplementary documentation, over a variety of tasks; Atwood and Ramsay (1978) found that a program design language was actually better than a flowchart for software design by graduate computer science students; Brooke and Duncan (1980a,b), who improved on the experimental techniques previously used, again found that flowcharts were of little help in debugging (though they did help in tracing execution flow) and that although they helped to localize the area where a bug was, they were insufficient to identify the bug.

Gilmore and Smith (1984) compared listings, flowcharts and structure diagrams as aids to debugging, and found no overall differences. Unlike most authors they were able to pursue the data analysis at a more detailed level, and found that subjects were using two main debugging strategies: some attempted to extract as much information as possible from program breakpoints, while others tried to form a mental model of the program. It appeared that among the subjects choosing the modelling strategy, listings – i.e. straight text, with no diagram at all – gave the fastest performance, while among the breakpoint-using subjects, flowcharts gave fastest performance. Thus, diagrammatic notations are likely to be good for certain purposes only. (Although even this picture was also affected by some individual differences between subjects and problems.)

Curtis *et al.* (1988) report systematic comparisons of several notations across several tasks using professional programmers. They compared three types of elements (flowchart symbols, pseudo-code, and natural language) arranged in three types of diagram (sequential listing, branching flowchart, or hierarchical construction), and studied the comprehension, coding, debugging, and modification of smallish programs. In the comprehension experiment they examined mental execution of programs; 'backwards' or 'deprogramming' problems, asking what input conditions must be met to achieve specified program behaviour; and comprehension of dataflow. Analysis of results from this large and well-designed study was unusually thorough and rigorous. Results indicated that the pseudo-code (or PDL, program design language) was best overall, although flowchart-like representations gave the best performance on tasks 'that accentuated the importance of tracing the control flow rather than grasping higher-order relationships'. The least-effective formats were those that employed natural language; 'in particular, sequentially presented natural language, the format of ordinary text, was especially ineffective on their tasks'. In general, the choice of element types carried more weight than the spatial arrangement. They also noted that individual differences accounted for more of the variance than any other variable! Later work (Boehm-Davis *et al.*, 1987) has shown that the reason for the superiority of the PDL may be that it lessens the 'translation distance' from the documentation format to the program code.

Despite these negative findings concerning flowcharts, Cunniff and Taylor (1987) and their co-workers have achieved reasonable success. Disturbed by student difficulties with introductory Pascal, they have devised a graphical language, FPL, which is a structured flowchart 'informationally equivalent' to Pascal. In one experiment they compared speed and accuracy of novices (drawn from a course teaching both FPL and Pascal in tandem) on recognition of simple structures, flow of control, and input/output, and evaluation (hand simulation) of simple program fragments. FPL was clearly superior. Note, however, that they did *not* use the 'deprogramming' tasks.

A further study (Cunniff *et al.*, 1989) has investigated novices' program construction. At present, although the data is scant, it seems that certain typical semantic errors are just as common in FPL as in Pascal (cf. the study by Spohrer *et al.* (1989) of novices' Pascal bug frequencies), but that certain other bugs appear to be rarer in FPL. The authors' explanation is a clear statement of the importance of making information accessible: FPL's 'spatial arrangement allows the user to think more clearly about the placement of such assignments [e.g. missing initializations]. This is especially helpful in the performance of updating and incrementing counters and running totals. The user is able to 'see' where the looping begins and ends, resulting in a more clear-cut definition of the correct location for updating' (p. 428). This explanation is, in fact, a statement that users must 'deprogram' their programs. Had the authors' previous study included a deprogramming task, it would have been very helpful to their argument.

FPL is an example of the structure diagram, of which a wide variety of different forms were critiqued by Green (1982). Some problems can be foreseen without empirical investigation. Take, for instance, the Nassi-Shneiderman diagram, which indicates that process B is a subcomponent of process A by writing the icon for B *inside* the icon for A: clearly the user will have to write smaller and smaller as the nesting gets deeper! Also some notations lent themselves better to subsequent modification than other notations, a vital requirement, as we saw above. However, there was – and still is – little empirical evidence available by which to make comparisons.

Recently, the hegemony of control-flow studies of traditional procedural notations has been challenged. Boehm-Davis and Fregly (1985), arguing that concurrent programming systems raised special problems, compared Petri nets to pseudocode and resource-sharing documentation. Swigger and Brazile (1989) compared times and errors for experienced subjects making modifications of two kinds (procedural and data oriented) to a rule-based expert system, supported by one of two types of diagram, entity relationship or Petri net, respectively expressing data relationships and order dependencies. Both these studies found that Petri nets, with their strong expression of control flow, were less successful, but both also concluded that the program size affected the issue and that further investigation of large programs would be needed. Regrettably, no determined effort has been reported to locate which tasks are best supported by each of these various types of documentation.

Bonar's work on BridgeTalk (Bonar and Liffick, 1990) offers a different alternative to control-flow, by working at the level of programming plans. (See Chapters 2.4, 3.1 and 3.2 for a description of programming plans.) Programs in BridgeTalk are constructed by slotting together plan components, and a simple set of subcomponents is used to show the data flow. The interleaved character of Pascal (not to mention of flowcharts), in which the initialization subcomponent of a plan may be located arbitrarily far from the 'focal line', is thereby replaced by a simple regime in which all the subcomponents are kept together, even at the price of simplifying the semantics. Other interesting decisions were made, which are not strictly at the level of information structure and will not be described here. This is virtually the only serious investigation of the *design* of a notation, rather than the comparison of existing notations; but, being motivated by educational objectives, it was not designed to yield general-purpose conclusions. Nevertheless, it brings out, better than the flowchart studies reviewed above, an important notational principle: ***strongly related subcomponents should be kept together, not dispersed.***

Unfortunately there are many more diagrammatic possibilities being touted by their supporters, than there are investigations.

4.5 Diagrams versus enhanced text

Instead of drawing diagrams and creating a new notation, one can continue to use the existing notation but use enhanced typography to make *perceptual cues reflect the notational structure.* There are obvious practical advantages, such as existing compilers, in sticking with existing notations. There are also more scientific advantages: it will help to clarify which claims about program visualization are true, and which are 'eyewash'.

A whole string of studies has now shown that ordinary program text can be made more comprehensible by supplying perceptual cues to important types of information. Payne *et al.* (1984) showed that a simple but hard-to-use command language for string editing was easier to use when commands were in upper case, argument strings in lower case. Isa *et al.* (1985) found that a meta-language for expressing syntax rules was more usable when structural cues were included. Gilmore (1986) showed that an improved version of Lisp 'pretty printing' helped his subjects both in writing and in comprehending programs, and Saariluoma and Sajaniemi (1989) showed that the components of spreadsheet programs were better recognized when the structure of the layout closely matched the internal structure of the spreadsheet, so that the perceived structure could be used to chunk the cells meaningfully.

In general, what ought to be useful is *improved access to the information that is (a) required, and (b) obscured.* The 'deprogramming' problem (Figure 1) illustrates one type of obscured information; 'programming plan' information is another type, since plans consist of statements that are dispersed in different places in the program. Gilmore and Green (1988) used colour cues, using the same colour for all components of the same plan, and showed that this one perceptual cue could improve learners' recognition of plan-based errors; their experiment also showed that control-flow error recognition was improved by indenting, and that cues to plans did not improve control-flow debugging, nor cues to control flow improve plan debugging. Thus the information-access hypothesis was well supported.

The Gilmore and Green results were obtained for small Pascal programs; Basic learners did not show the plan effect, but a later study by Davies (1990) showed that Basic programmers who had been taught structured programming techniques gave results similar to the Pascal programmers, while other Basic programmers thought mainly in terms of control flow – and were therefore presumably not seeking plan-like information. Van Laar (1989) has used the same technique to show that colour can supplement indentation in showing control flow in Pascal programs, with some net performance gain for learners answering a variety of comprehension questions. It would be extremely interesting to combine some of these techniques with the 'fish-eye' display (Furnas, 1986; see Chapter 1.2).

So the evidence is that the claims for improved comprehension from diagrams may be as well supported by non-diagrammatic techniques. The best conclusion at the moment, in the comparison between genuinely diagrammatic notations versus

```
 10 program prob12;
 20 vars depth, days, rainfall : integer;
 30   average : real;
 40 begin
 50   for days := 1 to 40 do
 60   begin
 70     depth := 0;
 80     writeln ('Noah, please enter todays rainfall');
 90     readln (rainfall);
100     rainfall := rainfall + depth;
110  end;
120  average := depth/40;
130  writeln('Average is', average);
140 end.
```

Figure 2: A small Pascal program from Gilmore and Green (1988). Shading represents the coloured highlighting of two plan structures, for forming a total in the variable depth and for inputting a value with a prompt. This program contains a 'plan' error at line 100 (it should read `depth := rainfall + depth`) and an 'interaction' error, combining plan and control-flow components, at line 70 (this line should be outside the For loop).

perceptual enhancements to text, seems to be that both are effective. There is no reason to suppose at present that diagrammatic notations are inherently superior. But it must be stressed that far too few of the diagrammatic possibilities have been properly investigated.

5 Programming as exploration

5.1 The implicit theory

We come now to the present view, which sees programming as opportunistic exploration or evolution, in which different alternatives are tried out and their consequences are considered. A favoured procedure is to find an old piece of program that 'almost' solves the present problem, and then to modify it bit by bit until the solution has been created. Recent languages and programming environments, led perhaps by Smalltalk, have put much emphasis both on re-usability and on the process of gradual, incremental change.

5.2 Research on change processes

We need to ask ourselves what features of notational and environmental design will support the process programming by placing fewest demands on the programmer. For learners, the tactical details of writing and changing code will be important. Gray and Anderson (1987) report the analysis of 'change episodes', points where a programmer (an advanced novice, in their study) altered the code that had been written. They argued that the most frequent changes would be made to those notational structures where the most planning was required, and further, that notational structures which could take a wider variety of forms would require more planning than structures that were more rigid. An example of such a plastic structure is the Lisp conditional structure. 'All conditionals have a Left parenthesis, a predicate, one or more clauses, and a Right parenthesis. Within this rigid structure the number, type, and the order of clauses can vary and it was these attributes that were involved in 15 change episodes. Because of the variability in how it can be used in Lisp, the goal-structure for the conditional ... must contain a large proportion of planning goals. The presence of planning goals is reflected by the frequency with which the conditional is involved in change-episodes' (p. 192). The authors contrast this with another form, the conditional clause, which is more rigid and which was involved only in the statistically expected number of change episodes.

As the authors point out, this study should not be over-interpreted, since it used only fifteen subjects and only a single notation: moreover, the fact that conditional structures cause problems was entirely predictable from earlier research. What they have done is to predict those problems from theoretical grounds, and to supply the firm prediction that notations with a more rigid, conventionalized structure would create fewer planning goals and therefore make programming easier.

Green *et al.* (1987) report a study scrutinizing performance in a variety of languages instead of only one, but paying correspondingly less attention to the fine details of individual behaviour. Using professional programmers working in their language of choice, they posed very simple problems (to reduce problem-solving behaviour to a minimum) and took detailed records of keystrokes as solutions were constructed. The key variable studied was the frequency of a 'backward move' to insert new material into part of the text that had already been written. The languages compared were Pascal, Basic and Prolog. Very similar solutions were produced in Pascal and Basic, but about four times as many backward moves were made in Pascal as in Basic, with Prolog occupying an intermediate position.

The most interesting aspects, from our viewpoint of notational design, came from considering the interplay of knowledge structures and code design. These authors accepted the 'programming plan' model of programmers' knowledge as a working hypothesis (see Chapters 3.1 and 3.2) in which plans contain a 'focal line' which achieves the goal of the plan. Sometimes a plan contains a precondition also, for example the counting plan, which contains a focal line `x := x + 1` and a precondition `x := 0`. They postulated that these preconditions would be forgotten more frequently than other plan components. Their data showed that the Pascal programmers more frequently went back to insert preconditions than the Basic programmers, and that the Prolog equivalent (the 'base case') was hardly ever the target of such a retrospective addition to the text. Could that be because base cases are spatially located right beside their associated focal line in the main case?

Davies (1989) extended this paradigm by examining the backward moves made by experts and intermediate learners in generating Pascal and Basic programs. The level of expertise was a more important discriminator than the programming language, but the most remarkable result was that intermediates jumped *within* plans while experts jumped *between* plans. Very similar results were found in a free-recall study. This takes us to questions of how knowledge representation interacts with notational structure, which we cannot consider here.

Research in this area is currently well behind the technology. Object-oriented programming languages are specifically designed with an information structure intended to support processes of change and modification, and especially of software reuse; yet we have very little evidence on whether they are successful – let alone on which of their many variations! Lange and Moher (1989) report a prolonged study of a single subject, who indeed frequently reused code, but – despite the intentions behind object-oriented programming languages – did so by copying text, rather than by using the special facilities of method-inheritance provided by the system. In general, 'the subject ... avoided techniques requiring deep understanding of code details or symbolic execution whenever possible'. While disappointing, no doubt, to the language designers, results like this do highlight the problems to be solved.

6 Lowering the barriers to programming

Studying how to improve programming notation has led us from considering whether some language features are intrinsically hard, through the problems of comprehension, into the question of how the notational structure interacts with the cognitive processes of planning and the 'fit' between the notational structure and the knowledge structure. As a side issue on the way we have examined the claim that visual notations are especially revealing, and have found little evidence to support it.

There is an urgent need for research on the interrelationship of notations and environments. To date, surprisingly little comparative work has been reported on environments, and virtually none that treats the problem of matching notation to environment. As it becomes easier to construct environments to support programming and programming-like activities, the possibilities will be explored and our knowledge will grow.

Knowledge structures and their interaction with notational design also need to be investigated in far greater detail than to date. Other chapters in this volume describe how our understanding of these topics is being investigated. No doubt, future studies will report further details of how notational designs can increase or decrease the need for local planning during coding, but also how the programmer's knowledge base can similarly affect the issue, in line with the work by Gray and Anderson reported above.

Where do we stand now? It seems that existing notations too often act to raise barriers against programming: barriers to learners, preventing understanding, and likewise to experts, preventing the fast access to information and modification that they need. There have been a few attempts to state design requirements for good languages, drawing on cognitive principles. Lewis and Olson (1987) suggest that, for the learner, one of the most important requirements is to suppress the 'inner world' of programming, the world of variable declarations, loops and input/output. The spreadsheet may be the model of the future, as they see it. (Although the information structure of ordinary spreadsheets has some problems – see Green, 1989).

Fitter and Green (1979) analysed the purported advantages of diagrammatic notations. Ten years later, I believe that their conclusions about notational design still hold good. Curtis (1989, p. 96) restated them so succinctly that I shall use his words:

> Fitter and Green (1979) argued that the primary problems in specification formats are the tractability and visibility of structure. Useful notations contained not only symbolic information, but also perceptual cues. For instance, maps use spatial location, histograms use variation in size, and Venn diagrams use spatial containment. They listed five attributes of a good notational scheme. These attributes are:
>
> (1) Relevance – highlights information useful to the reader.

(2) Restriction – the syntax prohibits the creation of disallowable expressions. [I.e. hard to understand, or likely to be confused with closely-related forms.]

(3) Redundant recoding – both perceptual and symbolic characteristics highlight information.

(4) Revelation – perceptually mimics the solution structure, and

(5) Revisability – easily revised when changes are made.

Of course, there is more to be said. But if programming notations met just those requirements, it would make a start.

Acknowledgements

My grateful thanks to Jean-Michel Hoc, Dorothy Graham, and especially Marian Petre, who all made detailed comments on an earlier version of this chapter.

References

Arblaster, A. T., Sime, M. E. and Green, T. R. G. (1979). Jumping to some purpose. *The Computer Journal,* **22,** 105-109.

Atwood, M. E. and Ramsay, H. R. (1978). *Cognitive structures in the comprehension and memory of computer programs: an investigation of computer debugging.* Alexandria, VA: U.S. Army Research Institute, TR78A21.

Boehm-Davis, D. A. and Fregly, A. M. (1985). Documentation of concurrent programs. *Human Factors,* **27,** 423-432.

Boehm-Davis, D. A., Sheppard, S. B. and Bailey, J. W. (1987). Program design languages: how much detail should they include? *International Journal of Man-Machine Studies,* **27,** 337-347.

Bonar, J. and Liffick, B. W. (1990). A visual programming language for novices. *In* S.-K. Chang (Ed.), *Principles of Visual Programming Systems.* Englewood Cliffs: Prentice-Hall.

Brooke, J. B. and Duncan, K. D. (1980a). An experimental study of flowcharts as an aid to identification of procedural faults. *Ergonomics,* **23,** 387-399.

Brooke, J. B. and Duncan, K. D. (1980b). Experimental studies of flowchart use at different stages of program debugging. *Ergonomics,* **23,** 1057-1091.

Chang, S.-K. (Ed.). (1990). *Principles of Visual Programming Systems.* Englewood Cliffs: Prentice-Hall.

Cunniff, N. and Taylor, R. P. (1987). Graphical versus textual representation: an empirical study of novices' program comprehension. *In* G. M. Olson, S. Sheppard and E. Soloway (Eds), *Empirical Studies of Programmers: Second Workshop.* Norwood, NJ: Ablex.

Cunniff, N. , Taylor, R. P. and Black, J. B. (1989). Does programming language affect the type of conceptual bugs in beginners' programs? A comparison of FPL and Pascal. *In* E. Soloway and J. C. Spohrer (Eds), *Studying the Novice Programmer.* Hillsdale, NJ: Erlbaum.

Curtis, B. (1989). Five paradigms in the psychology of programming. *In* M. Helander (Ed.), *Handbook of Human-Computer Interaction.* Amsterdam: Elsevier (North-Holland).

Curtis, B., Sheppard, S., Kruesi-Bailey, E., Bailey, J. and Boehm-Davis, D. (1988). Experimental evaluation of software documentation formats. *Journal of Systems and Software,* **9**, 1-41

Davies, S. P. (1989). Skill levels and strategic differences in plan comprehension and implementation in programming. *In* A. Sutcliffe and L. Macaulay (Eds), *People and Computers,* vol. V. Cambridge: Cambridge University Press.

Davies, S. P. (1990). The nature and development of programming plans. *International Journal of Man-Machine Studies,* **32**, 461-481

Fitter, M. J. and Green, T. R. G. (1979). When do diagrams make good computer languages? *International Journal of Man-Machine Studies,* **11**, 235-261.

Furnas, G. W. (1986). Generalized fish-eye views. *Proceedings of the CHI'86 Conference on Computer-Human Interaction.* New York: ACM.

Gilmore, D. J. (1986). Structural visibility and program comprehension. *In* M. D. Harrison and A. F. Monk (Eds), *People and Computers: Designing for Usability.* Cambridge: Cambridge University Press.

Gilmore, D. J. and Green, T. R. G. (1984). Comprehension and recall of miniature programs. *International Journal of Man-Machine Studies,* **21**, 31-48.

Gilmore, D. J. and Green, T. R. G. (1988). Programming plans and programming expertise. *Quarterly Journal of Experimental Psychology,* **40A**, 423-442.

Gilmore, D. J. and Smith, H. T. (1984). An investigation of the utility of flowcharts during computer program debugging. *International Journal of Man-Machine Studies,* **20**, 331-372.

Gray, W. and Anderson, J. R. (1987). Change-episodes in coding: when and how do programmers change their code? *In* G. M. Olson, S. Sheppard and E. Soloway (Eds), *Empirical Studies of Programmers: Second Workshop.* Norwood, NJ: Ablex.

Green, T. R. G. (1977). Conditional program statements and their comprehensibility to professional programmers. *Journal of Occupational Psychology,* **50**, 93-109.

Green, T. R. G. (1980). Programming as a cognitive activity. *In* H. T. Smith and T. R. G. Green (Eds), *Human Interaction with Computers.* New York: Academic Press.

Green, T. R. G. (1982). Pictures of programs and other processes, or how to do things with lines. *Behaviour and Information Technology,* **1**, 3-36.

Green, T. R. G. (1989). Cognitive dimensions of notations. *In* A. Sutcliffe and L. Macaulay (Eds), *People and Computers.* vol. V. Cambridge: Cambridge University Press.

Green, T. R. G., Bellamy, R. K. E. and Parker, J. M. (1987). Parsing-gnisrap: a model of device use. *In* G. M. Olson, S. Sheppard and E. Soloway (Eds), *Empirical Studies of Programmers: Second Workshop.* Norwood, NJ: Ablex.

Hoc, J.-M. (1981). Planning and direction of problem solving in structured programming: an empirical comparison between two methods. *International Journal of Man-Machine Studies,* **15,** 363-383.

Hoc, J.-M. (1988). Towards effective computer aids to planning in computer programming. *In* G.C. van der Veer, T.R.G. Green, J.-M. Hoc and D.M. Murray (Eds), *Working with Computers: Theory Versus Outcome.* London: Academic Press.

Holt, R. W., Boehm-Davis, D. A. and Shultz, A. C. (1987). Mental representations of programs for student and professional programmers. *In* G. M. Olson, S. Sheppard and E. Soloway (Eds), *Empirical Studies of Programmers: Second Workshop.* Norwood, N.J.: Ablex.

Isa, B. S., Evey, R. J., McVey, B. W. and Neal, A. S. (1985). An empirical comparison of two metalanguages. *International Journal of Man-Machine Studies,* **23,** 215-229.

Jones, C. (1979). A survey of programming design and specification techniques. *Specifications for Reliable Software.* IEEE Computer Society.

Lange, B. M. and Moher, T. G. (1989). Some strategies of reuse in an object-oriented programming environment. *Proceedings of the CHI'89 Conference on Computer-Human Interaction, 69-73.* New York: ACM, pp. 69-73

Lewis, C. and Olson, G.M. (1987). Can principles of cognition lower the barriers to programming? *In* G. M. Olson, S. Sheppard and E. Soloway (Eds), *Empirical Studies of Programmers: Second Workshop.* Norwood, NJ: Ablex.

Mayer, R. E. (1976). Comprehension as affected by structure of problem representation. *Memory and Cognition,* **44,** 249-255.

Myers, B. A. (1986). Visual programming, programming by example, and program visualization: a taxonomy. *Proceedings of the CHI'86 Conference on Computer-Human Interaction.* New York: ACM.

Naur, P. (1983). Program development studies based on diaries. *In* T.R.G. Green, S.J. Payne and G.C. van der Veer (Eds), *The Psychology of Computer Use.* London: Academic Press.

Payne, S. J., Sime, M. E. and Green, T. R. G. (1984). Perceptual structure cueing in a simple command language. *International Journal of Man-Machine Studies,* **21,** 19-29.

Pennington, N. (1987). Stimulus structures and mental representations in expert comprehension of computer programs. *Cognitive Psychology,* **19,** 295-341.

Saariluoma, P. and Sajaniemi, J. (1989). Visual information chunking in spreadsheet calculation. *International Journal of Man-Machine Studies,* **31,** 475-488.

Shneiderman, B. (1976). Exploratory experiments in programmer behavior. *International Journal of Information and Computer Sciences,* **52,** 123-143.

Shneiderman, B., Mayer, R. E., McKay, D. and Heller, P. (1977). Experimental investigations of the utility of detailed flowcharts in programming. *Communications of the ACM,* **20,** 373-381.

Shu, N. C. (1988). *Visual Programming.* New York: Van Nostrand Reinhold.

Sime, M. E., Green, T. R. G. and Guest, D. J. (1973). Psychological evaluation of two conditional constructions used in computer languages. *International Journal of Man-Machine Studies,* **5**, 105-113.

Sime, M. E., Green, T. R. G. and Guest, D. J. (1977). Scope marking in computer conditionals – a psychological evaluation. *International Journal of Man-Machine Studies,* **9**, 107-118.

Spohrer, J. C., Soloway, E. and Pope, E. (1989). A goal/plan analysis of buggy Pascal programs. *In* E. Soloway and J. C. Spohrer (Eds), *Studying the Novice Programmer.* Hillsdale, NJ: Erlbaum.

Swigger, K. M. and Brazile, R. P. (1989). Experimental comparison of design/ documentation formats for expert systems. *International Journal of Man-Machine Studies,* **31**, 47-60.

Van Laar, D. (1989). Evaluating a colour coding programming support tool. *In* A. Sutcliffe and L. Macaulay (Eds), *People and Computers V,* Cambridge: Cambridge University Press

Vessey, I. and Weber, R. (1984). Conditional statements and program coding: an experimental evaluation. *International Journal of Man-Machine Studies,* **21**, 161-190.

Youngs, E. A. (1974). Human errors in programming. *International Journal of Man-Machine Studies,* **6**, 361-376.

Wandke, H. (1988). User-defined macros in HCI: when are they applied? Paper delivered at *Macinter 2 Conference on Man-Computer Interaction.* Unpublished technical report from Sektion Psychologie der Humboldt-Universität zu Berlin, Oranienburger Str. 18, DDR-1020 Berlin, GDR.

Wetzenstein-Ollenschlaeger, E. and Schult, S. (1988). The influence of knowledge on the definition of procedures. Paper delivered at *Macinter 2 Conference on Man-Computer Interaction.* Unpublished technical report from Sektion Psychologie der Humboldt-Universität zu Berlin, Oranienburger Str. 18, DDR-1020 Berlin, GDR.

Chapter 2.3

Language Semantics, Mental Models and Analogy

Jean-Michel Hoc and Anh Nguyen-Xuan

CNRS - Université de Paris 8, URA 1297: Psychologie Cognitive du Traitement de l'Information Symbolique, 2, Rue de la Liberté, F-93526 Saint-Denis Cedex 2, France

Abstract

The semantics of a number of programming languages is related to the operation of a computer device. Learning a programming language is considered here from the point of view of learning the operating rules of the processing device that underlies the language, as a complement to the learning of new notations, or a new means of expression to be compared to natural language. This acquisition leads beginners to elaborate a new representation and processing system (RPS) by analogy with other systems that are associated to well-known devices. During acquisition, beginners not only learn new basic operations but also the constraints of these operations upon program structures. Learning therefore concerns a basic problem space as well as abstract problem spaces within which planning takes place. The links between this approach to learning to program and a number of related works on learning to use software are underlined. Implications of these research findings in the programmer training are drawn.

Psychology of Programming
ISBN 0-12-350772-3

1 Introduction

At first glance, programming may be defined as a procedure specification task by means of a computer language. This conception of programming, as pointed out by Miller (1974), led a number of researchers to stress the importance of the acquisition of the means – the computer language – in learning to program. Green offers an overview of this perspective in this book with his phrase 'dimensions of notations' (Chapter 2.2). Such a view is also prevalent in the research done by the Soloway team on Pascal (e.g. Bonar and Soloway, 1985). These researchers have established a taxonomy of errors done by novices in situations where they have been able to collect verbal protocols during program design. Their studies reveal the fact that many novice errors are due to wrong transfers of natural language constructs to computer programs (e.g. 'then' interpreted as 'afterwards' instead of 'in these conditions'). Confusion between Prolog notations of logic expressions and natural language or other formalisms are also shown to be an important factor of novices difficulties in learning Prolog (Taylor and du Boulay, 1986).

In teaching programming, the emphasis stressed on the means of expression can lead to an overestimation of the learning of programming language syntax within the triad determining human-computer interaction: external task structure, language syntax, and language semantics (Moran, 1981). Certainly a number of difficulties for beginners can be drastically reduced by designing ergonomic programming language syntax. For example, some interesting properties of a two-level syntax have been demonstrated (task-action grammar: Payne and Green, 1986). This enables the user to learn a limited number of general rules that play the role of schemas and which can be instantiated so that several specific rules can be generated. Nevertheless these syntactic schemas have their semantic counterpart and the efficiency of this kind of syntax is probably determined by a certain compatibility with a pre-existing two-level structuration of the contents of the language.

Syntactic errors have been shown to be of limited importance even to beginners (Youngs, 1974). In a number of studies on learning to program, which stress the acquisition of a new communication means, semantic rather than syntactic difficulties are shown. This is especially true of Soloway's works which show that beginners introduce distortions into programming language syntax when their programming knowledge is lacking. These distortions are indicators of transfers from other knowledge domains that are not compatible with the programming language structure. Even with professional programmers, this kind of transfer, from a well-known programming language to a new one, has been shown to exist (Hoc, 1988b). Syntactic errors clearly revealed semantic difficulties when programmers did not succeed in transforming well-known contents into quite different contents expressible in the new language. This phenomenon often occurs in translation between natural languages. Taylor and du Boulay (1987) show the Prolog experts' ability to adopt a problem representation compatible with the language in the very beginning of the design process. Within the same study, programmers who are expert in programming but unfamiliar with Prolog initiate their design activity with representations that are not compatible with Prolog and very often fail to produce a program.

In this chapter, the relationship between task structure and programming language semantics will be discussed as a critical component of learning to program. Semantics and syntax are considered to be complementary components in the study of user models. Although a major effort has been made to bridge the gap between natural language and programming languages (mainly for English-speaking programmers) the semantic problem remains. As Jackson (1980) has pointed out, it is almost impossible to design languages that are purely problem oriented. They remain largely machine oriented and their semantics consist of controlling machine operation (especially a sequential mode of operating in performing tasks where human operators may use parallel processing). Indeed, this machine is a formal one and varies with the language used: e.g. the Pascal machine is different from the Cobol machine. Research is only beginning on more 'declarative' languages such as Prolog but the need to learn the operating rules of the Prolog machine is already shown to be a necessary condition to designing complex programs in this kind of language (Taylor and du Boulay, 1987).

The purpose of this chapter is to show that whatever the programming language, beginners have to learn the operating rules of what is called the 'device' underlying the programming language. This learning develops along two directions:

(1) construction of new elementary operations (primitives), different from familiar ones, which are sometimes coded by the same wordings (especially the READ and WRITE statements, as has very often been shown);

(2) restructuring of well-known plans which are incompatible with the new primitives – beginners become aware of the constraints upon the structure of the new plans defined by these new primitives.

Beginners learn to program by building programs. Hence they learn by problem-solving. The important features of this problem solving situation will be presented here, giving reasons for the theoretical framework used in analysing the elaboration of user models of language and devices: learning by doing and by analogy. A basic construct will then be introduced (after Hoc, 1977, 1988a), in relation to mental models, to define the knowledge architecture within which these learning mechanisms take place: the representation and processing system (RPS). General problem solving by analogy strategies will be discussed in the context of human computer interaction studies aimed at describing the beginner's acquisition of computerized tasks. And finally, some implications on training will be stressed.

Mainly research on the learning of procedural languages will be referred to, since it is available in books or journals. Similar components can be found in the learning of more declarative and recent languages as shown by studies that have not been widely published.

2 Problem solving by beginners in programming

Whatever the kind of learner and whatever the teaching method may be (e.g. top-down structured programming methods, or algorithmics), programming language acquisition remains the first necessary step to more advanced acquisitions.

As has been stressed above, learning to program is learning by problem solving. A number of studies have been devoted to beginner problem solving strategies in these learning-by-doing situations and it is now possible to define their principal features. After Hoc (1988a), the notion of problem is to be contrasted against the notion of task. A task is defined by a goal and conditions for reaching it. A problem is a representation of a task a subject evokes or elaborates that cannot yet trigger an acceptable procedure to reach the goal (i.e. a procedure which is in conformity with the prescribed conditions).

Beginners are confronted with computer tasks, like stock updating, file sorting, etc., for which they have no available procedures. These tasks cannot be considered as problems for professional programmers who can activate well-known computer schemas ready to fulfil them (see Chapter 3.1). The notion of problem is therefore subjective: it is related to the interaction between subject and task and must be defined by both the task characteristics and the subject's knowledge. What then constitutes a programming problem for a beginner in contrast to an expert?

More often than not teachers do not ask beginners to invent algorithms. Beginners are required to produce programs for familiar tasks they can perform by hand. In other words the goals are familiar to them and they have procedures at their disposal so that these goals can be reached. These familiar tasks become problems because there are mainly two new conditions that must be satisfied:

(1) Programs have to be practicable by a computer through a definite means of communication: the programming language (this communication situation is very different from the human-human communication situations where interpretation takes place). Programmers thus have to restrict themselves to the use of elementary operations the computer can perform and of procedure structures compatible with these operations.

(2) New procedures have to be explicitly stated beforehand (in most of the cases new elaborations cannot take place at the execution time); hence a strong planning constraint is introduced.

In this chapter the first kind of condition will be discussed, the second one is examined elsewhere in this book (see Chapter 2.4). However, some short comments on the constraints on procedure expression must be given here. To be able to express procedures in a program, beginners must elaborate representations of these procedures, such representations may not be necessary in usual problem-solving situations. In ordinary situations the procedure can be elaborated at execution time, where concrete feedback is available after the execution of each operation: in programming, feedback is delayed. Usually the goal is specific (e.g. sorting of a particular file for which a particular procedure can be adopted) while in most of the programming situations the goal is a class of specific goals (e.g. the sorting procedure must be valid for any file defined by a set of characteristics). This creates two kinds of difficulty for beginners:

* shift from value to variable processing,

* elaboration of a representation of the procedure control structures of which beginners are not necessarily aware in usual problem solving situations.

In several experiments, we have adopted a methodology whereby the two aspects of problem solving in beginners – the design of a procedure practicable by the computer, and the expression of this procedure – are studied separately (Hoc, 1983; Nguyen-Xuan and Hoc, 1987). Beginners are first asked to perform a task by hand in a situation with as few constraints as possible and the familiar procedure is observed. They are then asked to command a computer device to carry out the task step by step: the elementary commands available correspond to elementary instructions of a programming language. In this command situation all the data to be processed (e.g. an entire file) and the content of the computer memory cells are visible in a first stage and covered in a second stage. These two situations enable the experimenter to observe the adaptation of the familiar procedures to the operating rules of the device and force the beginners to adopt a general procedure (covered situation). Finally the beginners are asked to verbally state their general procedure after having elaborated it in the command situations where they had feedback information (step-by-step error messages in the covered situations). Figure 1 illustrates the application of this kind of methodology to the sorting of a list.

This methodology enables the observer to identify the diverse sources of difficulties that beginners encounter when designing a program in standard programming situations: acquisition of the operating rules of the new device and procedure expression.

3 The concept of representation and processing system

In human-computer interaction as well as process control areas the notion of the mental model has been introduced to describe operator knowledge about machines (Rouse and Morris, 1986). Although the uses of this concept are various, the common idea is to describe a kind of operational knowledge that is specific to a limited class of situations. The application of mental models to operator behaviour raises methodological difficulties that will not be discussed here. In particular these mental models must be opposed to 'conceptualizations' as observer constructs (Norman, 1983). Although conceptualizations can only be defined by the observer as compatible with expert behaviour (instead of being actually used by the operator), explicit teaching of these conceptualizations in learning to use a device can provide the learner with a helpful representation of the device, and can be of interest (du Boulay *et al.*, 1981).

In a study of the learning of programming, Hoc (1977) has introduced the notion of the representation and processing system (RPS) which is similar to a 'mental model'. A RPS is a part of the semantic memory network that can be activated in executing tasks belonging to a common task domain (see Hoc, 1988a, for further developments). Such a network is elaborated by instruction and by doing, and has the double status of a social and individual construct. As different task domains are separately taught – algebra, French, physics, etc. – they are separately internalized in the form of different RPSs. Connections between pieces of knowledge belonging to the same RPS are stronger than connections between different RPSs.

In a RPS, declarative (representation) and procedural (processing) knowledge are strongly connected as dual aspects of knowledge. This accounts for the fact that formally isomorphic tasks may not be processed in the same way, because they are represented differently by the subject. A particular kind of task representation opens access to certain procedural skills that cannot be so easily triggered by another representation. It may happen that the same problem statement triggers different RPSs for solving various sub-problems, and the communication between different RPSs may be difficult to manage. In an experiment conducted by Hoc (1977), subjects at different levels of programming expertise had to solve a problem of ticket

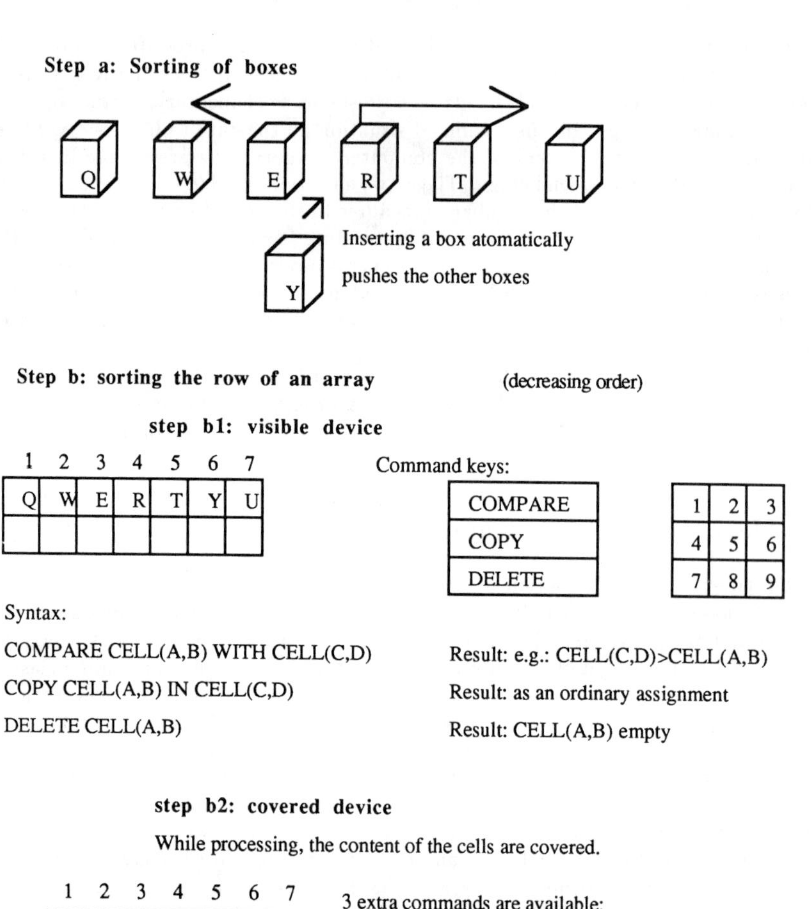

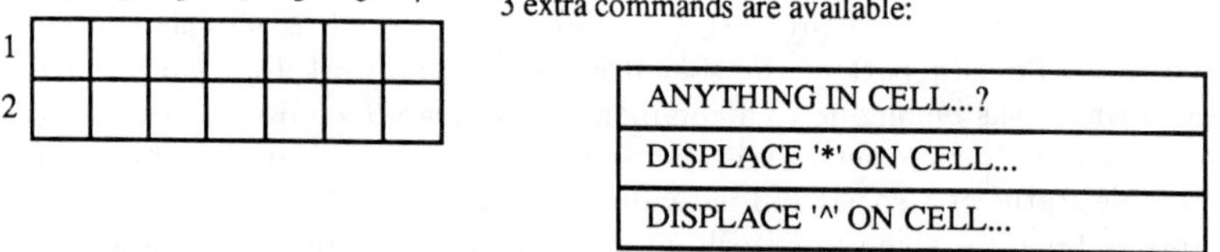

In trying to transfer the box insertion procedure,

a shifting sub-procedure has to be applied before

inserting a letter.

Figure 1: Sorting of a list (after Nguyen-Xuan and Hoc, 1987). (a) Familiar situation: sorting of boxes. (b) Sorting of a row in an array: (b1) visible contents, (b2) covered contents.

machine control simulation in a Metro station. In solving this kind of problem, a number of difficulties encountered by subjects were representation and processing translations from four separate RPSs into a common 'computer' RPS:

* 'traveller', in which traveller goals (e.g. ticket asked) and actions (coins inserted) were represented;

* 'ticket machine', describing the machine reactions to the traveller and computer messages;

* 'numerical code', the transfer of information and commands between the computer and the ticket-machine;

* 'accounting', the knowledge domain in which the relation between information and commands took its meaning (e.g. computation of the change to be returned).

Given the multiplicity of task domains that have been internalized by an individual, a number of RPSs are available and are triggered by problem statements or perceptual cues. These knowledge structures play a central role in learning new domains and dealing with novel tasks. They are the source of problem solving by analogy; for example in learning computerized tasks – triggering of a RPS related to the typewriter domain in learning to use a text editor (Wærn, 1989), or to the calculator in learning to use a turtle command language (Shrager and Klahr, 1986: calculator 'view application').

RPS can be defined at several levels of abstraction; from a basic definition of elementary operations (e.g. the operation meant by a read statement), and data properties or relations (e.g. an integer variable), to abstract operations (e.g. iteration to go through a file) and representations (e.g. file structure). This hierarchy enables individuals to plan their actions (see Hoc, 1988a, for further developments about this hierarchy of abstract spaces). The basic definition is called the (basic) device associated to the RPS, the precision of which depends upon the individual's expertise and goals. For example the Pascal machine can be considered as the basic device in designing programs, but a more elementary device would be considered if compiler or execution error messages had to be understood.

Research into use of computer devices offers a comparison between two types of knowledge that are embedded in the same RPS (Hoc, 1978; Young, 1981; Richard, 1983):

* operating rules that describe the conditions of validity of the operations and their effects, and can only be 'surrogates' (Young, 1981) without causal semantics or based on deeper knowledge;

* utilization rules that describe operations to be used in relation to goals, and make task-action mapping easier.

Each of these types of knowledge can be defined at different levels of abstraction. Operating rules can not only refer to very elementary operations of the device but also to macro-operations. Utilization rules may reflect different levels of analysis of situations, as has been shown in different contexts (Card *et al.*, 1983; Wærn, 1989; Richard, 1986):

* the goal (or task) level describing the intentional or functional aspects of the activity;

* the means (or method) level at which procedural aspects of goal attainment are processed;

* and the prerequisite (or condition) level giving access to details such as those that are necessary to process interactions between elementary operations.

As far as operating rules are concerned, it has been shown that the levels of abstraction correspond to processing priorities, from the analysis of goals down to the analysis of prerequisites (Richard, 1986; Morais and Visser, 1987). So, if a goal representation can trigger a familiar procedure, this procedure is applied to a novel device before detecting mismatches which lead to a deeper analysis of the situation.

4 Problem solving by analogy in programming

4.1 Rationale for an analogical transfer

We argue above that people learn to program through practice, i.e. they learn by problem solving. But unlike learning a knowledge domain (e.g. physics), the learner knows a procedure in a RPS associated to a well known device (e.g. a box-sorting procedure by hand, although the task has to be accomplished by a computer with an array: see Figure 1). That is, a means is available for reaching an analogue goal state from an analogue initial state. The problem consists in making the computer reach the goal.

We suggest that in such a situation the learner is prone to relying on a familiar RPS to build the goal structure needed, instead of elaborating a new goal structure compatible with the computer from scratch. In other words, the learner will consider that the computer solution is analogous to the familiar one. But goal structures available within this RPS are compatible with the device associated to the RPS without being necessarily compatible with the device underlying the programming language. So some more-or-less profound adaptation will take place before reaching an acceptable program.

4.2 Empirical evidence

Hoc (1977) has shown that, in learning procedural programming, beginners who are dealing with tasks they can execute by hand may evoke procedural plans that are available in familiar RPS. They then try to refine these plans until reaching the level of the programming language statements: in most cases, this refinement leads either to a detection of incompatibilities between the plans and the device, or to reach programs which are not optimal. This phenomenon is reinforced in training beginners to use top-down programming methodology before learning the operating rules of the computer device (Hoc, 1983). The incompatibilities between these plans and the device involve mainly sequential data access and result production modes different from those used by the human cognitive system which is capable of parallel processing.

Most of the beginner programming errors can be interpreted as wrong analogical transfers from other RPS to use of the computer device (Bonar and Soloway, 1985).

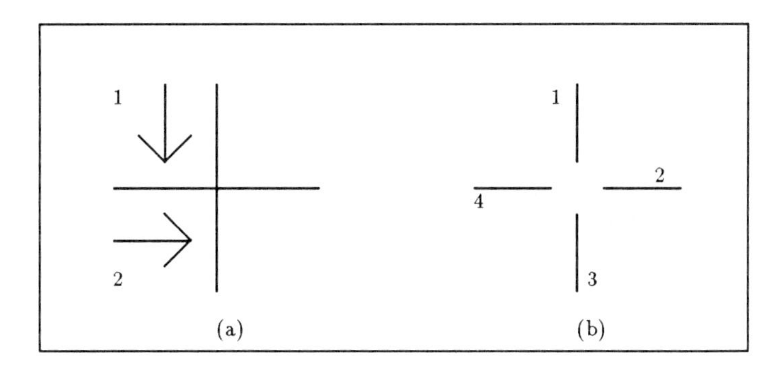

Figure 2: The two representations of the cross (after Mendelsohn, 1986): (a) two orthogonal lines (before LOGO learning), (b) four orthogonal lines (after LOGO learning).

Moreover, in executing a task where several methods are available, the program is designed from the method most frequently executed by hand. This has been shown in an experiment on sorting: beginners tried at first to adapt the sorting-by-insertion method they readily used by hand, even if it was more difficult to program than the extremum method (Nguyen-Xuan and Hoc, 1987).

In learning a new programming language, the RPS elaborated to program in an already familiar programming language can be used as an analogue. Van Someren (1984) points out that when starting to program in Prolog beginners who already know an instruction-oriented language just try to implement an algorithm and then translate it into Prolog rules.

Indeed, analogical transfer is a paramount phenomenon. Beginners have been shown to borrow plans from their execution-by-hand RPS. But experienced programmers hand-manipulate files in a different way from novices: they use the plans they would have used in a computer program (Eisenstadt *et al.*, 1985). This 'reverse' analogical transfer may concern the structure of the representation of well-known objects which has been modified in order to be adapted to computer programming. Mendelsohn (1986) has found such transformations in elementary school pupils who learn Logo. For example, the cross which is genuinely seen as composed of two orthogonal lines is seen after the Logo course as composed of four orthogonal lines (Figure 2).

The importance of analogical transfers at the outset of new device learning has been stressed by a number of authors in very different contexts: text editing (Wærn, 1985; Allwood and Eliasson, 1987), pocket calculator use (Bayman and Mayer, 1984; Friemel and Richard, 1987). When analogical transfer is impossible from well-known RPSs this transfer may come from previously acquired constructs in the same language as has been shown in learning Lisp (Anderson *et al.*, 1984) or in operating a robot (Klahr and Dunbar, 1988).

4.3 Effects of analogy

More often than not authors stress the negative effects of the analogical transfer when trying to interpret novice errors. However, two kinds of analogical transfer must be distinguished: a transfer through an abstract schema, the effect of which can be positive, and a transfer by direct mapping, which has been frequently shown to be confusing.

An experiment on learning to use electronic devices – such as notepad, clock, or chequebook – conducted by Kamouri *et al.* (1986) clearly demonstrates the superiority of an exploration-based training over an instruction-based training, thanks to the inducement of analogical reasoning in the exploration situation. This experiment, however, uses a well-suited frame to obtain positive effects from analogy, and follows Gick and Holyoak's works (1980, 1983) which reveal in particular that the condition for an efficient analogical transfer to occur is a prior elaboration of an abstract schema. This elaboration of an abstract schema has been shown to be improved when subjects are required to solve several analogous problems. So Kamouri *et al.* had their subjects solve problems with three analogous devices before discovering a new device either analogous to the previously acquired ones or not. The authors insist that this kind of situation is difficult to find in real work settings. But it may be a possible explanation of the well-known fact that learning a new programming language or software is improved by previous knowledge of other similar programming languages or software.

In the human-computer interaction area, analogical transfer by direct mapping is very often observed in beginners: two analogous objects are confused instead of being considered as different instantiations of the same schema. For example, the text editor space bar is considered to have exactly the same function as the typewriter space bar. These confusions are reinforced by the choice of familiar command names or programming language statements similar to natural language descriptions of actions. If feedback is available, this kind of direct transfer can stimulate active learning, for it can trigger accommodation processes, and differences are as useful as similarities (Carroll and Mack, 1985).

4.4 Mechanisms of learning by analogy

Several research studies shed light on the mechanisms of acquisition of operating rules in diverse learning-by-doing situations and on the conditions of efficiency of these situations. In experimental game situations where subjects are mostly unfamiliar with the games (Anzai and Simon, 1979; Nguyen-Xuan and Grumbach, 1985), four main mechanisms have been described:

* progressive elaboration of the problem space;

* identification of wrong actions which lead to errors, and generation of procedures in order to avoid them;

* identification of correct actions which lead to the goal, and creation of subgoals to which these actions can be applied;

* structuring of the subgoals by processing goal interactions.

But the story could be somewhat different in learning to program, and, more generally, in learning to use a command device.

4.4.1 Borrowing problem space and goal structure

In learning to use software or to program, the problem space is not generated from scratch, but rather from familiar problem spaces relating to familiar devices (e.g. typewriter when using and editor) or RPSs that are related to the problem domains (e.g. algebra, management, accountancy, etc., when programming). In the case of programming, familiar goal structures are transferred from familiar RPS as a number of studies have shown (Hoc, 1977, 1983; Nguyen-Xuan and Hoc, 1987). Hence, in contrast to usual learning by solving problems, the novice programmer already has a goal structure at the outset, although it may not be quite relevant. Sometimes, this goal structure comes from other programs (examples used in programming courses or handbooks, programs written by anybody else, etc.).

In Hoc's experiment (1983) on stock updating by the mean of a computer device, novices tried to use the '+' command corresponding to the ordinary binary addition as a summation operation after having entered a list of numbers to be added into the same memory cell. In our experiment on sorting (*op. cit.*) we observed a clear transfer of the insertion method used by all the subjects in the execution-by-hand situation. The goal structure of this method consists in decomposing the problem into iterative subgoals. Each one corresponds to the insertion of a new element into a well-sorted sub series. This goal structure was transferred into the command device situation, although it was very costly in comparison to the constraints of the device. Most of the subjects tried to locally modify subgoals, instead of building a totally new goal structure.

4.4.2 Repairing goal structure

When a goal structure is available, attempts are made to reach the subgoals by means of the device operators. A number of difficulties have been pointed out, which lead the learner to modifying the borrowed goal structure (Carbonell, 1983; Nguyen-Xuan, 1987). Some of them are particularly relevant to the use of a new computer device:

* The necessity to satisfy preconditions in the target situation, which are automatically satisfied in the source situation. For example, in our experiment on sorting (*op. cit.*) the subjects transferred the insertion method. In the source situation where boxes are sorted, the insertion action automatically pushes the adjacent boxes to make room. In the target situation, where an array is used, 'room made' is a precondition that must be satisfied by a quite complex shifting procedure (see Figure 1).

* The necessity to decompose elementary source actions into even more elementary ones. The already cited example of assimilating the '+' operator to a summation is illustrative of this decomposition.

* The discovery of unexpected goal interactions that are not present in the source procedure. An example can be found in Burstein's work on learning Basic (1986). Some novices assimilate the assignment statement to stacking in a box. Interaction occurs when they want to store a series of values: each new value deletes the previous one, which does not happen in the case of stacking in a box.

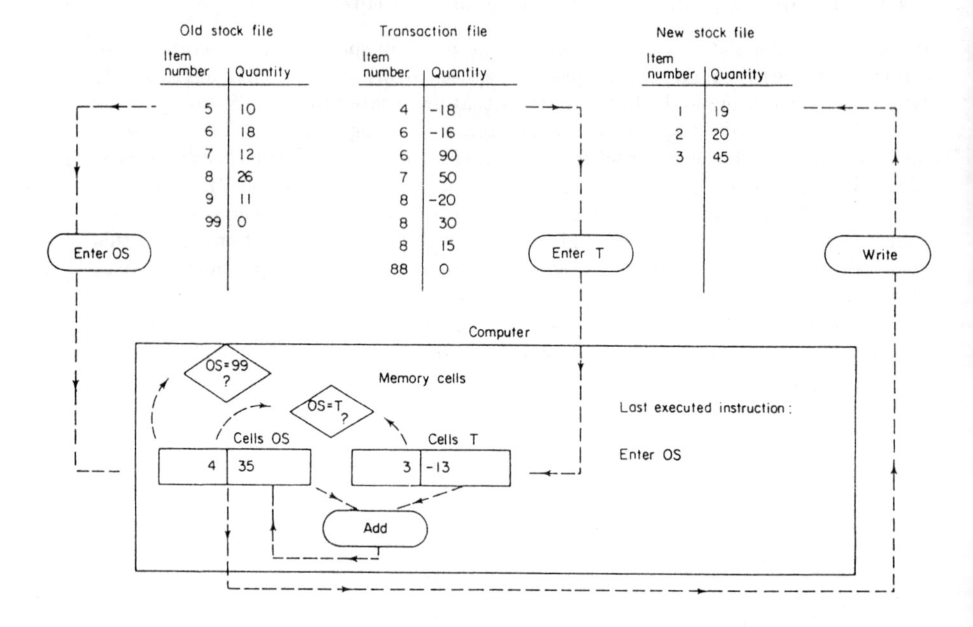

Figure 3: Updating device (after Hoc, 1983): (a) visible situation, subjects can see the contents of the files and they visually compute the number of transactions before entering an item from the old stock file; (b) covered situation, they have only access to the contents of the memory cells and cannot identify the number of transactions before processing an item.

* The incompatibility between the source goal structure and the data identification means. In a procedure, not only transformation operations but also identification operations must be considered. Identifications concern data properties that are relevant for choosing the appropriate action to perform (e.g. in an updating task, the identification of the number of transaction per item). The available means to perform the identifications may deeply affect the goal structure. This is the reason for difficulties encountered by novices when they had to transform a 'read-process' iterative schema into a 'process-read' one (Soloway *et al.*, 1982; Samurçay, 1985) as is shown by an experiment conducted by Hoc (1983).

In the above-mentioned experiment by Hoc, novices had, successively, to update a stock file either with visual access to the files or with access to only one item and one transaction at a time in computer memory cells (Figure 3). In the former situation the identification of the type of item (number of transactions) was performed before processing the item, so the 'read transaction - process it' schema was used. In the latter this identification was no longer possible, so the 'process transaction – read the next one' schema had to be used to identify the end of the processing of the current item. The subjects showed very strong reluctance to abandon the 'read-process' goal structure: they tried to stick to it by attempting to perform actions that were not allowed. When they noticed that the transaction that had just been entered did not correspond to the item being processed, some subjects tried to return the transaction to the input file so they could enter it when they processed the item to which it corresponded.

5 Implications for training

From the results presented above, some implications on training design can be drawn, which could facilitate the elaboration of an operative device model and lead to the generation of optimal procedures. In most of the cases, the learner relies too heavily on available goal structures which may be inadequate, following the processing priority of the goal analysis level and neglecting the other levels: means and prerequisite analysis (see above).

(a) Learning situations have to be designed to prevent the learners from only being oriented towards the attainment of the task goal.

Friemel and Richard (1987) noticed that the presentation of a visual simulation of the internal operation of a pocket calculator was not efficient enough to enable the acquisition of operating rules. Novices devoted more attention to attaining of the goal than to analysing the display. This led the authors to ask novices to predict the effects of written procedures before learning by doing. This method has proved to be efficient when different procedures leading to the same result are presented: the novices are then encouraged to turn from a goal-level analysis to a means-level analysis. Certain recommendations proposed by Wærn (1989) lead in the same direction: encouraging the learner to use various methods of performing the same task, to process unexpected results, and to reflect on observations.

(b) Goal structures which could be transferred have to be known by the teacher so that learning situations can be designed, which force the learner to abandon these structures or which facilitate the adaptation.

In the first case constraints can be imposed which reveal the inadequacy of the source goal structure. In our experiment on sorting (*op. cit.*) the subjects had a two-dimensional array at their disposal (see Figure 1). The elements to be sorted were displayed on the first line and the well sorted series could be constructed on the second line. In this situation most of the subjects adhered to their insertion method without being able to discover an optimal procedure. However, when the subjects were given a single line and an extra cell (transfer) they rapidly turned from the insertion procedure to an optimal extremum procedure. In the second case the question concerns the effectiveness of the error identification and the possibility of recovering from errors. This principle underlies the recommendation (Carroll and Carrither, 1984) to design successive devices which from the onset restrict degrees of freedom in behaviour so that the feedback should be easy to manage.

(c) Even analogical transfer by direct mapping may be helpful, but with the condition that a sufficient amount of feedback is provided to the learner.

Immediate feedback has been shown to be more effective than delayed feedback. In an experiment on learning to use a pocket calculator Friemel and Richard (1987) found that an exploration-based situation is more efficient than a situation where subjects are encouraged to program their procedures before executing them. Immediate feedback is more effective, probably because it enables subjects to causally connect elementary actions to their direct effects. This has been shown in an experiment where naive subjects had to elaborate a sorting procedure in a stepwise command situation (Nguyen-Xuan and Hoc, 1987). At the beginning of the experiment subjects were given access to the content of the cells of a table they had to sort (see Figure 1). They then had to do the same task without access to the content of cells. No subject succeeded in changing his/her procedure to an optimal one in the covered situation: the change, when it occurred, was made in the visible situation.

Unfortunately, in a programming situation, one has to write an entire program before being able to get feedback from the execution. In addition, error messages are not as easily understood as they should be. The aim of these messages is to protect the compiler or the operating system rather than operative evaluation of the error (du Boulay and Matthew, 1984). Nevertheless, the absence of feedback for the beginners may be somewhat repaired by the presentation of a concrete model of the computer device. Several experiments conducted by Mayer (1975, 1976, 1981) have shown that the presentation of a device model prior to learning did have a positive effect on program interpretation, iteration management, and transfer to novel situations, in contrast to the sole presentation of the language rules. However, these effects are not found neither in program generation nor when subjects are good at mathematics. These results are interpreted by the author in the context of meaningful learning: the model enables subjects to assimilate new pieces of knowledge with previously acquired knowledge.

6 Conclusion

Research into the acquisition of mental models of computer devices shows how important it is to consider the programming language as a semiotic tool, the content of which corresponds to the operation of a device. This acquisition implies learning-by-analogy mechanisms which cannot be successful without analysis of the feedback obtained from execution.

Learning-by-doing situations, however, have been shown to present limitations if they are not designed to improve spontaneous mechanisms. The crucial improvement consists in discouraging the learner to concentrate on goal-level analysis of the task and encouraging him to access to means level and prerequisite level. In addition the learner has to be assisted in identifying and recovering from errors. Investigation into learning-by-doing mechanisms points towards design of this kind of support.

Most of the research works concerning the very start of learning to program have mainly been concerned with procedural languages. Research on more recent programming languages such as Prolog or object-oriented languages (Smalltalk) are now in progress. Results concerning Prolog (Bundy *et al.*, 1986; Taylor and du Boulay, 1986; White, 1988) suggest that the story could be somewhat different. Prolog has some special features (backtracking, unification, pattern-directed control, etc.) that cannot be found in other procedural languages or everyday situations. But although the Prolog user has to deal with logic specification, the learning of the Prolog machine is necessary to the implementation of the specification. Procedural aspects of Prolog programs are shown to play an important role even at an expert level, where tracing tools are very often used (Taylor and du Boulay, 1987). Results on object-oriented languages are not yet available.

This construction of a mental model of device operating rules mainly applies to the start of the learning process. It results in the acquisition of elementary rules as well as goal structures compatible with the novel device. Macro-operators and programming plans, which contain the operation constraints of the device as built-in constructs, are learnt. Complex procedures can then be designed without analysing the details of the device operation. Certainly these learning-by-doing situations have to be restricted to the elaboration of simple procedures. The stepwise command of complex procedures (e.g. with deep iteration embedding) would lead to a mental load too heavy to be practicable.

The subsequent stages of learning to program, through the acquisition of adequate problem representations, programming plans, and schemas, are examined in the next chapter of this book.

References

Allwood, C.M. and Eliasson, M. (1987). Analogy and other sources of difficulty in novices' very first text editing. *International Journal of Man-Machine Studies*, **27**, 1-22.

Anderson, J.R., Farrell, R. and Sauers, R. (1984). Learning to program in LISP. *Cognitive Science*, **8**, 87-129.

Anzai, Y. and Simon, H.A. (1979). The theory of learning by doing. *Psychological Review*, **86**, 124-140.

Bayman, P. and Mayer, R.E. (1984). Instructional manipulation of user's mental models for electronic calculators. *International Journal of Man-Machine Studies*, **20**, 189-199.

Bonar, J. and Soloway, E. (1985). Preprogramming knowledge: a major source of misconceptions in novice programmers. *Human-Computer Interaction*, **1**, 133-161.

Bundy, A., Pain, H., Brna, P. and Lynch, L. (1986). *A proposed PROLOG story.* University of Edinburgh, Department of Artificial Intelligence, Research paper no 283.

Burstein, M.H. (1986). Concept formation by incremental analogical reasoning and debugging. *In* R.S. Michalski, J.G. Carbonell and T. Mitchell (Eds), *Machine Learning: An Artificial Intelligence Approach*, vol. 2. Los Altos, CA: Kaufmann.

Carbonell, J.G. (1983). Learning by analogy: formulating and generalizing plans from past experience. *In* R.S. Michalski, J.G. Carbonell and T.M. Mitchell (Eds), *Machine Learning.* Palo Alto, CA: Tioga, pp. 137-161.

Card, S.K., Moran, T.P. and Newell, A. (1983). *The Psychology of Human-Computer Interaction.* Hillsdale, NJ: Erlbaum.

Carroll, J.M and Carrithers, C. (1984). Training wheels in a user interface. *Communications of the ACM,* **27,** 800-806.

Carroll, J.M. and Mack, R.L. (1985). Metaphor, computing systems, and active learning. *International Journal of Man-Machine Studies,* **22,** 39-57.

du Boulay, B. and Matthew, I. (1984). Fatal errors in pass zero: how not to confuse novices. *In* G. van der Veer, M.J. Tauber, T.R.G. Green and P. Gorny (Eds), *Readings on Cognitive Ergonomics – Mind and Computers.* Berlin: Springer-Verlag, pp. 132-143.

du Boulay, B., O'Shea, T. and Monk, J. (1981). The black box inside the glass box: presenting computing concepts to novices. *International Journal of Man-Machine Studies,* **14,** 237-249.

Eisenstadt, M., Breuker, J. and Evertsz, R. (1985). A cognitive account of 'natural' looping constructs. *In* B. Schackel (Ed.), *Human-Computer Interaction – INTERACT 84.* Amsterdam: North-Holland, pp. 455-459.

Friemel, E. and Richard, J.F. (1987). Apprentissage de l'utilisation d'une calculette. *Psychologie Française,* **32,** 227-236.

Gick, M.L. and Holyoak, K.J. (1980). Analogical problem solving. *Cognitive Psychology,* **12,** 306-355.

Gick, M.L. and Holyoak, K.J. (1983). Schema induction and analogical transfer. *Cognitive Psychology,* **15,** 1-38.

Hoc, J.-M. (1977). Role of mental representation in learning a programming language. *International Journal of Man-Machine Studies,* **9,** 87-105.

Hoc, J.-M. (1978). La programmation comme situation de résolution de problème. Paris, Université René-Descartes, Doctoral Dissertation.

Hoc, J.-M. (1983). Analysis of beginner's problem-solving strategies in programming. *In* T.R.G. Green, S.J. Payne and G. van der Veer (Eds), *The Psychology of Computer Use.* London: Academic Press, pp. 143-158.

Hoc, J.-M. (1988a). *Cognitive Psychology of Planning.* London: Academic Press.

Hoc, J.-M. (1988b). Towards effective computer aids to planning in computer programming. *In* G. van der Veer, T.R.G. Green, J.M. Hoc and D. Murray (Eds), *Working with Computers: Theory Versus Outcomes.* London: Academic Press.

Jackson, M. (1980). The design of conventional programming languages. *In* H.T. Smith and T.R.G. Green (Eds), *Human Interaction with Computers.* London: Academic Press, pp. 321-347.

Kamouri, A.L., Kamouri, J. and Smith, K.H. (1986). Training by exploration: facilitating the transfer of procedural knowledge through analogical reasoning. *International Journal of Man-Machine Studies,* **24,** 171-192.

Klahr, D. and Dunbar, K. (1988). Dual space search during scientific reasoning. *Cognitive Science,* **12,** 1-48.

Mayer, R.E. (1975). Different problem-solving competencies established in learning computer programming with and without meaningful models. *Journal of Educational Psychology,* **67,** 725-734.

Mayer, R.E. (1976). Some conditions of meaningful learning for computer programming: advanced organizers and subject control of frame order. *Journal of Educational Psychology,* **68**, 143-150.

Mayer, R.E. (1981). The psychology of how novices learn computer programming. *ACM Computing Surveys,* **13**, 121-141.

Mendelsohn, P. (1986). Activation de schèmes de programmation et mémorisation de figures géométriques. *European Journal of Psychology of Education,* **1**, 126-138.

Miller, L.A. (1974). Programming by non programmers. *International Journal of Man-Machine Studies,* **6**, 237-260.

Morais, A. and Visser, W. (1987). Programmation d'automates industriels: adaptation par des débutants d'une méthode de spécification de procédures automatisées. *Psychologie Française,* **32**, 253-259.

Moran, T.P. (1981). The command language grammar: a representation for the user interface of interactive computer systems. *International Journal of Man-Machine Studies,* **15**, 3-50.

Nguyen-Xuan, A. (1987). Apprentissage par l'action d'un domaine de connaissance et apprentissage par l'action du fonctionnement d'un dispositif de commande. *Psychologie Française,* **32**, 237-246.

Nguyen-Xuan, A. and Grumbach, A. (1985). A model of learning by solving problems with elementary reasoning abilities. *In* G. d'Ydewalle (Ed.), *Cognition, Information Processing, and Motivation.* Amsterdam: North-Holland.

Nguyen-Xuan, A. and Hoc, J.-M. (1987). Learning to use a command device. *European Bulletin of Cognitive Psychology,* **7**, 5-31.

Norman, D. (1983). Some observations on mental models. *In* D. Gentner and A.L. Stevens (Eds), *Mental Models.* Hillsdale, NJ: Erlbaum, pp. 7-14.

Payne, S.J. and Green, T.R.G. (1986). Task-Action Grammars: a model of mental representation of task languages. *Human-Computer Interaction,* **2**, 93-133.

Richard, J.F. (1983). Logique du fonctionnement et logique de l'utilisation. Le Chesnay (F), INRIA, Research Report No. 202.

Richard, J.F. (1986). The semantics of action: its processing as a function of the task. Le Chesnay (F), INRIA, Research Report No. 542.

Rouse, W.B. and Morris, N.M. (1986). On looking into the black box: prospects and limits in the search for mental models. *Psychological Bulletin,* **100**, 349-363.

Samurçay, R. (1985). Learning programming: an analysis of looping strategies used by beginning students. *For the Learning of Mathematics,* **5**, 37-43.

Shrager, J. and Klahr, D. (1986). Instructionless learning about a complex device: the paradigm and observations. *International Journal of Man-Machine Studies,* **25**, 153-189.

Soloway, E., Ehrlich, K., Bonar, J. and Greenspan, J. (1982). What do novices know about programming? *In* A.Badre and B. Shneiderman (Eds), *Directions in Human-Computer Interactions.* Norwood, NJ: Ablex.

Taylor, J. and du Boulay, B. (1986). Why novices may find programming in Prolog hard? University of Sussex, Cognitive Studies Research Paper No. 60.

Taylor, J. and du Boulay, B. (1987). Learning and using Prolog: an empirical investigation. University of Sussex, Cognitive Studies Research Paper No. 90.

Van Someren, M. (1984). Misconceptions of beginning Prolog programmers. University of Amsterdam, Department of Experimental Psychology, Memorandum 30.

Wærn, Y. (1985). Learning computerized tasks as related to prior task knowledge. *International Journal of Man-Machine Studies,* **22**, 441-455.

Wærn, Y. (1989). *Cognitive Aspects of Computer Supported Tasks.* Chichester: Wiley.

White, R. (1988). Effects of Pascal upon the learning of Prolog: an initial study. University of Edinburgh, Working paper.

Young, R.M. (1981). The machine inside the machine users' models of pocket calculators. *International Journal of Man-Machine Studies,* **15**, 51-85.

Youngs, E.A. (1974). Human errors in programming. *International Journal of Man-Machine Studies,* **6**, 361-376.

Chapter 2.4

Acquisition of Programming Knowledge and Skills

Janine Rogalski and Renan Samurçay

CNRS - Université de Paris 8, URA 1297: 'Psychologie Cognitive du Traitement de l'Information Symbolique' 2, Rue de la Liberté, F-93526 Saint-Denis Cedex 2, France

Abstract

Acquiring and developing knowledge about programming is a highly complex process. This chapter presents a framework for the analysis of programming. It serves as a backdrop for a discussion of findings on learning. Studies in the field and pedagogical work both indicate that the processing dimension involved in programming acquisition is mastered best. The representation dimension related to data structuring and problem modelling is the 'poor relation' of programming tasks. This reflects the current emphasis on the computational programming paradigm, linked to dynamic mental models.

1 Introduction

Learning to program in any language is not an easy task, and programming teachers are well aware of the myriad difficulties that beset beginners. Why is it so difficult to learn or teach programming? How can analysis of these difficulties serve to indicate

Psychology of Programming
ISBN 0-12-350772-3

ways of remedying them? The process of teaching and learning not only involves learners but a set of situations in which teachers stage knowledge about programming. Teachers need to know more than the mechanisms of human learning in order to analyse and structure their teaching. They also must be aware of the fact that what learners learn is dependent on their own conception of the activity of programming, and choices and decisions they as teachers make during instruction.

This chapter presents a framework for the analysis of programming. It serves as a backdrop for a discussion of findings on programming learning and is a basis for an understanding of why some programming concepts and procedures are difficult to teach and to learn. This framework is based on a conceptual view of programming. Programming, like problem solving activities in other scientific fields such as physics or mathematics, can be analysed in terms of expert competence (expertise-oriented framework), or in terms of a constituted knowledge domain (content-oriented framework). Knowledge of this latter type is socially and historically constructed, and is formed of normative and symbolic representations used for communicative purposes.

Studies in the field of programming learning can be categorized by their orientation. A number are expertise oriented, where a model of expert knowledge is used as a reference to analyse novices' errors and misconceptions (Bonar and Soloway, 1985; Soloway and Ehrlich, 1984). Other studies are content oriented, in that greater attention is paid to an analysis of programming knowledge as a domain having its own concepts, procedures, notations and tools (Pea; Rogalski; Rouchier, Samuçay; Taylor and du Boulay). A third category of studies takes no explicit epistemological position on the analysis of programming knowledge as a specific task domain. Programming is used here primarily as a paradigm which lends itself to the testing of general models of subjects' cognitive architectures and learning processes (Anderson *et al.*, 1984).

This chapter focuses on the acquisition of programming knowledge, as testified to by students' ability to solve 'programming problems' at various levels of complexity. Acquisition of programming skills for general educational goals and its developmental aspects are discussed in Chapter 2.5. The students referred to here are adults or adolescents with general educational backgrounds although they may be complete novices with respect to programming knowledge. Thus we will be looking at issues of computer literacy and those related to pre-professional and professional training.

We will be arguing that the teaching process cannot reproduce the real process of construction of knowledge by scientific or professional people in the classroom. Rather the teaching process creates new 'objects for teaching' which need to satisfy certain functional properties in order to create meaning in learners' minds. Pair's work (see Chapter 1.1) is used as a springboard for showing that the impact on the teaching process of teachers' conceptualizations of programming activity leads to the development of different types of knowledge and strategies in learners.

2 A framework for programming activity

2.1 General framework

Figure 1 presents a general theoretical framework for knowledge representation in the programming field. It is made up of four related 'spaces': the knowledge structure, problem solving, practice and cognitive tools. This framework is illustrative of the fact that knowledge is constructed and assessed by problem solving. The knowledge

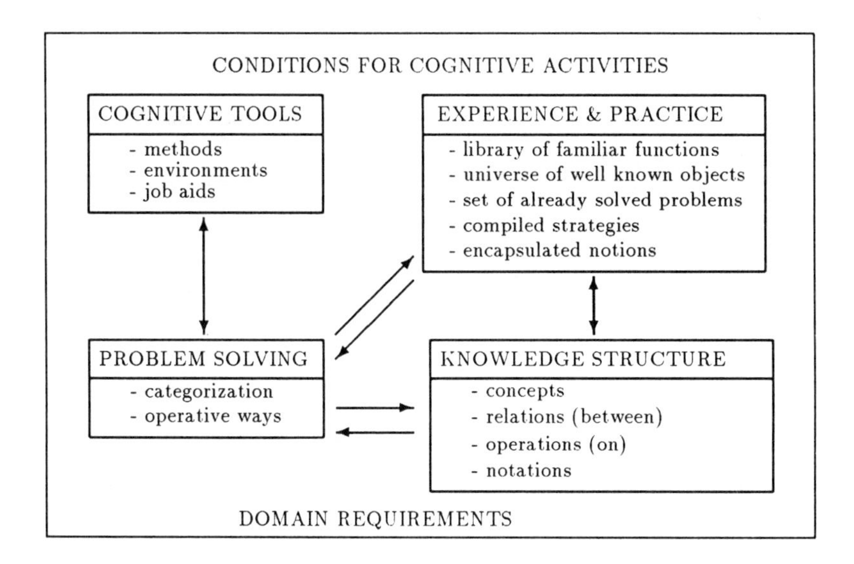

Figure 1: Frame for task analysis.

structure consists of a hierarchical system of concepts, operations on, and relations between concepts having a set of notations. In any problem in the field, concepts need to be co-ordinated. For instance a list problem involves variables, iteration or recursion, list functions, test, inputs and outputs, whatever is the programming language, and whatever are its semantical and syntactical specificities. Appropriate notation needs to be selected (even within a programming language, choices must be made between different control structures). In addition there is a relationship between types of problems and operative pathways. Sum problems, sort problems, list or graph processing are related to different classical algorithms. Cognitive activities in programming tap both prior experience and individuals' cognitive 'tool bag'.

During the process of knowledge acquisition, all four 'spaces' evolve. New notions are acquired through new interactions. For instance, the parameter passing in procedures or functions involves the notion of procedure and a redefinition of variables as global or local, and thus generates a new level of understanding of the relationship between variable, name and value.

Complex data structures gradually become familiar objects (such as tables for beginners in procedural languages for example). The co-ordination of functions or procedures can be conceptualized in terms of the role they play in the problem, without reference to mental running with specific values (encapsulated notion of functions). The set of previously solved problems increases. Certain strategies become directly available, relating types of problems with operative pathways (compiled strategies). The complexity of problems that can be tackled increases.

At the start of the acquisition process students necessarily refer to previous everyday experiences, the acquisition of the first RPS (representation and processing

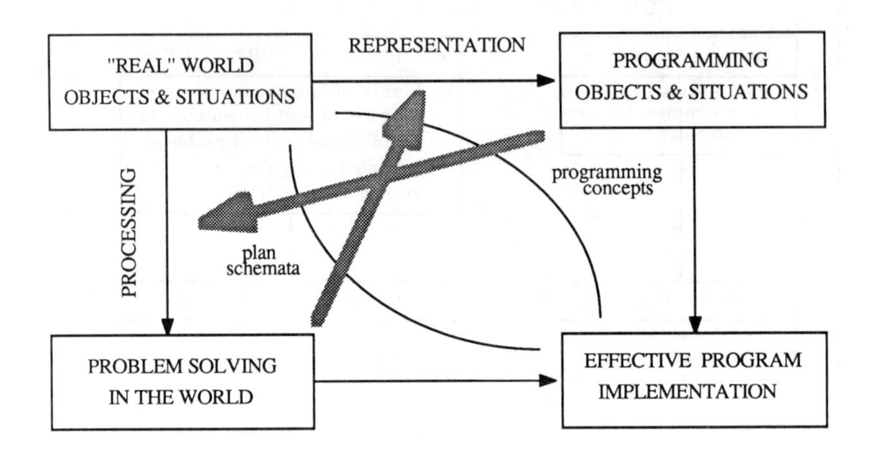

Figure 2: The task of programming.

system, see definition in Chapter 2.3) can be built up by such mechanisms as learning by doing, or by analogy, or from other fields of knowledge such as mathematics (the prime source), physics, or from exposure to other programming languages. Precursors of this type can generate both negative as well as positive effects because of similarities and differences in knowledge structures (Rogalski and Vergnaud, 1987). These effects may account for two types of empirical results. Preprogramming knowledge is known to be a source of misconceptions (Bonar and Soloway, 1985), and previous programming knowledge can even constitute an obstacle, requiring an 'unlearning process' (Taylor and du Boulay, 1987; Chapter 4.2). Also, students with mathematical backgrounds do not need the same amount of teaching (van der Veer and van der Wolde, 1983; van der Veer *et al.*, 1986).

2.2 What is programming

Figure 2 presents a schema for analysing the cognitive activities involved in programming tasks 'from a real-world problem to a runnable program text'. In certain respects this schema is the psychological counterpart of the epistemological analysis developed in Chapter 1.1.

The crucial dimensions in the activity of programming are processing and representation. There are two ways individuals can move from a real-world problem to program text implementable on a given device. A real-world problem can first be solved in the domain and then translated into program text. Or, alternatively, it can be approached in the programming language and applied to the real-world object. When problem solving takes place on a real-world object, processing precedes representation, even if the properties of the objects in the target programming language intervene in the choice of solution. This approach is closely related to the 'computational' programming paradigm where a program is defined as a succession of computations. The 'functional' perspective where the program is seen as a function that needs to be decomposed is more infrequent. When problem solving initiates in the programming text, the structuring of data and relationships between pieces of data is the core of programming activity.

Studies on the acquisition of programming skills are mainly centred on the processing dimension. There are at least three reasons for this: (1) the historical role of procedural languages, (2) the importance of planning in programming design (Hoc, 1988a) and (3) the productive role of organization of actions as a first programming model (Samurçay and Rouchier, 1985). In addition, problems given to novices are often directly defined in terms of programming entities, such as numerical data and variables in the well-known sum problems (Spohrer *et al.*, 1985; Soloway *et al.*, 1982; Soloway and Erhlich, 1984; Samurçay, 1985) or explicitly in terms of a given programming language, such as: 'define a function called list-sum. Given a list of numbers, list-sum returns the sum of these numbers. For example, list-sum′ (5 10 − 4 27) returns 38. (List-sum′ ()) returns 0' (Katz and Anderson, 1988).

Few studies have dealt with the representation task related to data structuring even though a number have reported findings on the effects of problem content or semantics (Ormerod *et al.*, 1988; Widowski and Eyfert, 1986). In procedural languages, complex data structures are often related to complex problems and the representation task may only appear crucial for advanced students. However, there is evidence that program design aids may cause difficulties when the language is insufficiently rich in data structures flexible enough to represent the variety of data (Hoc, 1988b). In a relational language like Prolog, beginners may run up against two difficulties: choosing how to represent objects and the relationship between them, and deciding how general the solution should be (Taylor and du Boulay, 1986). From the cognitive analysis developed by Rosson and Alpert (1988), it can be hypothesized that the same difficulties could be encountered with object-oriented programming languages.

The acquisition of programming knowledge and skills can be characterized in a number of ways. Problem solving in programming can be centred on problems in the real world (research on planning) or on the program as text (research on programming language and use). Here, schemas are defined as sets of organized knowledge used in information processing, and plans are defined as organized sets of dynamic procedures related to static schemas. For a given problem, plans and schemas can be defined at several levels (strategy, tactics, implementation) (Samurçay, 1987).

Studies on the construction and instantiation by experts and novices of schemas (the program-as-text perspective) or plans (the programming-as-an-activity perspective) can be categorized as a function of these perspectives. Program text studies are centred on the cognitive activities involved in the understanding or debugging of written programs (see Chapter 3.1). In these experiments schemas are defined as standard structures that can be used to achieve small-scale goals such as those involved in sum problems using variable plans and loop plans. Training research indicates that schemas can be induced by creating analogies between training problems and programs, but are most efficient in understanding and debugging tasks, and less efficient in program design.

We take a slightly different point of view which is more oriented towards the acquisition of fundamental programming concepts that can be implemented in various programming tasks (design, execute, modify or complete programs). One reason for this emphasis is that bugs in novice programming are mainly conceptual errors (Spohrer and Soloway, 1986). In the framework presented above (Figure 1) the development of strategies is seen as that facet of acquisition related to problem solving, and is more general than either plan or schema acquisition (see also Chapter 3.2). There are advantages in seeing plans and schemas as special cases of 'compiled' strategies and 'encapsulated' notions, related to the set of previously solved problems: these concepts are valid whatever programming language is concerned (Rosson and Alpert, 1988). Methods (related to task analysis) may enhance the cognitive activities involved in strategy research and management and help programmers in problem solving.

3 Cognitive difficulties in learning programming

Programming as a knowledge domain differs from other neighbour domains such as mathematics or physics in two ways. First, there are no everyday intellectual activities that can form the basis for spontaneous construction of mental models of programming concepts such as recursion or variables, in contrast to such notions as number or velocity. These concepts, however, are the basis for concept acquisition in programming. Secondly, programming activity operates on a physical machine which may not be transparent in its functioning for learners.

Which familiar RPS can be activated in learners for the construction of a new RPS that is operational for programming? Which kinds of transformations are needed for the construction of this new RPS? How do beginners construct the conceptual invariants related to programming?

The difficulties encountered by novice programmers have been subdivided below into four areas corresponding to the main conceptual fields novices must acquire during the learning process. This subdivision has been made purely for readability's sake since it is obvious that novices suffer from syndromes rather than from single misconceptions. All errors arise in conjunction with others and may have multiple causes as a function of the problem context.

3.1 Conceptual representations about the computer device

When designing, understanding, or debugging a program, expert programmers or software developers refer necessarily to their knowledge of the systems underlying programming languages, and the programming tools available on these systems. They are able to move from one system to another and change their representation of the problem to adapt to new constraints (see current practices of professional programmers in Chapter 4.2). This knowledge extends beyond operating rules and involves the representation of the whole system (editor, operating system, input/output devices, etc.) This is what Taylor and du Boulay have termed the 'notional machine'.

There are two basic difficulties encountered by novices as concerns conceptual representations. The first involves the effects of the relative complexity of the command device in the acquisition process. The second is related to the construction of the notional machine.

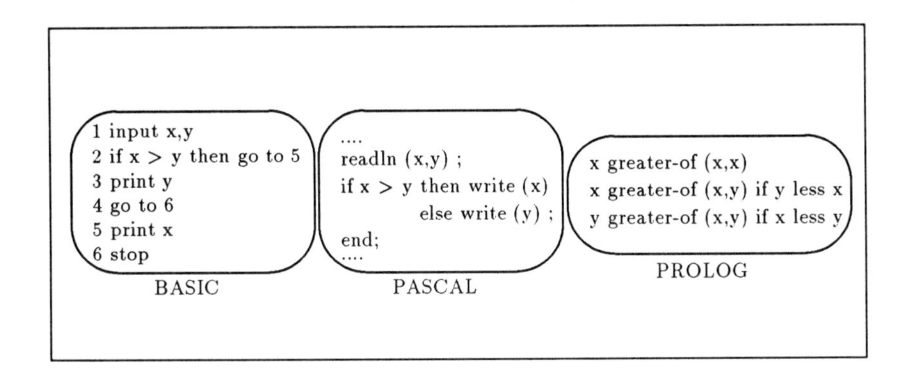

Figure 3: Examples of programs in Basic, Pascal and Prolog.

The very elementary level of construction of a RPS by beginners for simple command devices has been described in Chapter 2.3. This first RPS constructed 'by doing' and by analogy is, however, not sufficient for full conceptualization of the notional machine with respect to the command device. The notion of 'complexity' of a command device is itself a highly relative concept. Learning a text editor, how to use a pocket calculator, or a programming language are not analogous tasks because of the features inherent to the problems each device can be used to solve. Even in programming languages, the conceptualization of a 'Basic machine' a 'Pascal machine' or a 'Prolog machine' will not involve the same cognitive operations. For instance, compare the programs in Figure 3 written in these three languages that will print the larger of two input numbers.

Basic and Pascal programs describe *how* a result is computed; they specify actions to be performed by the machine. The Prolog program describes *what* is true of the problem solution, leaving it up to the computer to sort out the steps involved in finding the solution. In the case of imperative languages, a first, although extremely limited RPS can be built on the representation of *actions* and their sequential control. In languages like Prolog this representation requires the concept of *logical truth* and a representation of how Prolog *evaluates* this truth. Thus the RPS that beginners must construct with Prolog involves objects such as truth values and relations that are conceptually far removed from the objects involved in imperative languages. Moreover, as Taylor and du Boulay (1987) point out, mastery of these objects in the pure logic domain is not enough. The way Prolog processes to evaluate the truth of predicates (backtracking) has no counterpart in 'manual' logic.

There are two dimensions to the relative complexity of the command device: the proximity of the control, and the existence of virtual entities (virtual memories, variables, files, etc.) simulating entities which have no physical identity. It is believed that the command device increases in complexity for learners as a function of the distance of the control and the variety of virtual entities. For example, assembler language and Prolog constitute extremes in this case. Prolog is used here for purposes of illustration of a language that differs from imperative ones and identical analysis could be made of object-oriented languages such as Smalltalk. It would be worth-

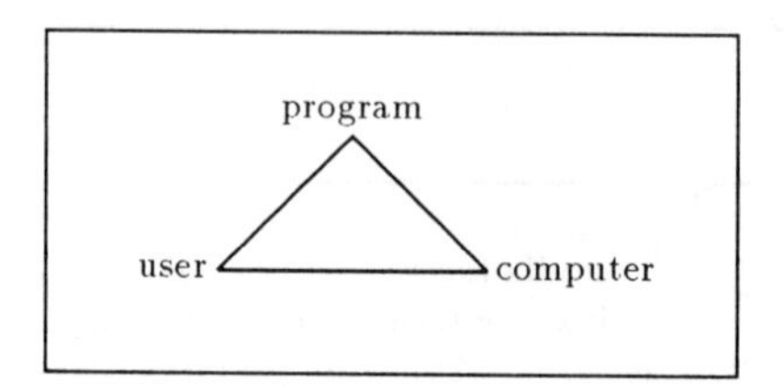

Figure 4: Relationships involved in program running.

while to investigate how experts deal with virtual entities, which are an important component of 'high-level' programming languages.

In program design, beginners also need to construct mental models of the functional relationships between program, user (who will enter the data) and the computer device the program is run on (Figure 4).

Misrepresentations of these relationships interfere with planning activity in beginners. They also affect the meaning beginners assign to the results of execution and to error messages. Novices' erroneous representations of the notional machine can be analysed from two points of view.

The first concerns language-independent conceptual bugs which disrupt the way in which novices program and understand programs. Several terms have been coined to describe these misconceptions: 'superbugs' (Pea, 1986), 'meta-analysis' (Hook *et al.*, 1988), 'preprogramming knowledge' (Bonar and Soloway, 1985), 'wrong RPS transfer' (Hoc and Nguyen-Xuan, Chapter 2.3). These mainly characterize errors in communication rules involving computer jumps into the real-world domain, or false expectations that the program will flow from top to bottom and left to right.

A number of studies have focused on these errors. Rogalski and Hé (1989) present a model ('PRES') implemented by novices when dealing with conditional problems in a Basic-like language. In this model novices assume that 'in a conditional part of a program when condition A has been processed the other instruction is applied in the notA situation (it is unnecessary to specify the latter condition notA in the program text); conversely as long as the processing of situation A is not terminated, the instruction should be applied to case A '. This model illustrates an inadequate representation of sequentiality and is based on natural communication rules. It, however, applies successfully to CASE-OF problems (where there is no sequentiality problem) but fails with NEST-style conditional problems.

Effects of misconceptions have also been observed in simple iterative programs designed by novices (Laborde *et al.*, 1985; Samurçay and Rouchier, 1985; Bonar and Soloway, 1985). A common error consists of writing the operations in the order 'description of actions/repeat mark' which again reflects the fact that communication with the computer is considered to be an instance of natural language communication. Even when beginners acknowledge that there is no mind inside the computer they still credit it with semantic abilities including presupposition and interpretation of the content of communication with the user.

Similar errors have also been reported for Prolog novices on tracing problems, or analysis of the result of trace functions. Novices tend to jump into the real-world domain in order to solve the problem; this model works in restricted circumstances (meaningful variable names, simple and familiar problem domain) but fails when the problem cannot be solved without a minimal understanding of the backtracking mechanism (Hook *et al.*, 1988). Novices may also introduce certain features of their reading skills, for instance by assuming that control flow through a clause or through a program always goes from left to right (Taylor and du Boulay, 1987).

The second type of error related to representations of the notional machine has to do with the system as a whole. Novices not only have problems with the representation of the machine underlying programming languages but also encounter difficulties with the computing system they are working with. Beginners need to differentiate which elements of this system belong to the language, and which are system entities. Prolog novices frequently alter their programs and forget to reload or reconsult the new file; alternatively they may inadvertently assert what are meant to be queries, thereby accidentally altering the program/database (Taylor and du Boulay, 1987). They have comparable difficulties maintaining the distinction between the Prolog program/database they can see on the terminal screen and the version of the program/database Prolog will process. Even in imperative languages where the distinction is clearer between program and data, novices do not discriminate clearly 'who' is controlling the input/output commands when the program is running, or the functional role of outputs on the screen with respect to memory management (Rogalski and Samurçay, 1986).

3.2 Control structures

One of the major points made by Pair (Chapter 1.1) is that programming can be conceptualized in a variety of ways including describing calculations (computational paradigm), describing functions (functional paradigm), or defining and processing relations between objects (relational paradigm). Regardless of the type of conceptualization or type of language, programs must be able to describe: (1) an undefined number of processings which remain to be described in a finite manner; (2) a general solution for a set of infinite data. Conditional and iterative or recursive structures are built as responses to these requirements in various forms depending on the language. The prime characteristic of control structures is the fact that they disrupt the linearity of the program text.

Control structures can be analysed and taught in different ways as a function of conceptualization. They can be seen (and taught) as *computational models* (oriented towards the description of calculations in the program), or as *functional models* (oriented towards analysis of the function of the program with respect to the problem). This distinction does not entirely mirror the declarative versus the procedural distinction since objects as well as processings are involved in both.

In the majority of studies describing early acquisition processes, as well as in programming tutoring systems, the computational approach dominates. Few studies have examined the acquisition of programming from the functional point of view. Research on the acquisition of control structures by novices has mainly centred on conditionals, iterative and recursive statements, and procedural languages. There is no explicit conditional statement in Prolog, and conditional treatments have to be handled via logical conditions (such as conjunctions, negation, etc.), and backtracking controlling (such as 'cut'). The problems novices encounter in the conceptualization of backtracking is illustrative of the level of difficulty they face in the conceptualization of control structures (Taylor and du Boulay, 1986, 1987).

3.2.1 Conditional statements

The earliest studies in this field examined the elaboration of programming languages and compared programming styles such as JUMP style using the GOTO statement, NEST style using the embedded alternatives IF .. THEN .. ELSE, CASE-OF style using a succession of positive conditions (IF...THEN). Overall these studies indicate that the JUMP style, although seemingly easier in terms of expression, is more difficult to control than the NEST style (Green, 1980) and that styles using positive alternatives present fewer difficulties for acquisition. In addition, syntactic markers such as BEGIN..END defining the scope of the THENs and ELSEs and spatial organization of the text make conditional structures easier to learn and use (Sime *et al.*, 1977). However, difficulties in understanding conditional statements are also related to the type of question, which can be either conditional or sequential (Green, 1980), and the organization of logical cases. Depth of nesting of conditionals also increases difficulty. This set of studies also shows that there is no uniform 'mental language' used by novices in acquiring conditional structures (Gilmore and Green, 1984).

In terms of the acquisition of notions, logical and mathematical precursors have been shown to play an important role. At a global level, students with more grounding in mathematics learn new structures more rapidly (van der Veer *et al.*, 1986; van der Veer and van der Wolde, 1983). The ability to use logical connectors (AND, OR, and NOT) and represent structured cases are necessary for the acquisition of conditionals, but they are not sufficient in themselves because of the role played by an inappropriate model of sequentiality of execution by the 'computer device' (Rogalski and Hé, 1989).

3.2.2 Iteration

The construction of an iterative plan involves the identification of elementary actions/rules which must be repeated, and the condition governing end or continuation of the repetition. There are three operations entering into the construction of an iterative plan: construction and expression of the loop invariant (updating), identification of the end control and its place in the loop plan (test), and identification of the initial state of the variables (initialization).

There are two types of iterative problems. In the first, the end control is a constant, whereas in the second the end control is a value of a variable computed in the loop. In the latter instance, two plans are possible with respect to goal attainment

process variable/test variable

or

test variable/process variable

These two forms are not equally accessible in a the novice's existing plan catalogue (Soloway *et al.*, 1982; Samurçay, 1985). Novices' 'spontaneous' iterative models have the following structure: description of actions, repetition counter, repetition mark and expression of the end control. Anticipation and explicitation of the end control are not spontaneously available (Laborde *et al.*, 1985; Samurçay and Rouchier, 1985), probably because this control is implicit in ordinary action plans. These features make it more difficult for novices to deal with 'test/process' plans (e.g. the WHILE loop in Pascal). Difficulty stems from representing and expressing a condition about an object on which they have not yet operated (Soloway *et al.*, 1982).

A didactic study shows that in a set of teaching situations where a variety of sum problems involving different types of constraints were introduced, novices could successfully be guided in the construction of an adequate representation of different loop plans (Samurçay, 1986). However, these representations are unstable and fail on new problems; novices return to step-by-step construction. In other words, the novice model is based on a representation of a succession of actions (dynamic model) rather than on a representation of the invariant relationship between the different states of variables (static model).

A typical loop plan implemented by novice programmers is given below:

```
................
sum := 0 + number
counter := 1
sum := number1 + number2
counter := 2
repeat
.................
```

This fragment of protocol reveals two types of difficulties. The first is the construction of the loop invariant (`sum := sum + number` and `counter := counter + 1`) and the second is related to the designation of variables. The novice tends to use different names at each step in execution to label the same functional variable.

As indicated above, most research work, as well as tutoring systems such as Bridge (Bonar, 1984) or Proust (Johnson and Soloway, 1983), on the acquisition of the iterative control structure has been conducted from a computational point of view. It is likely that computational teaching reinforces the novices' dynamic model. This model fails when novices encounter new problems because they have not acquired how to search and construct the 'loop invariant' (relations must be conserved between variables during execution). The representation of the succession of variables and specific training on search for relationships between them could have positive impact on the learning process. Even with object-oriented language, it was observed difficulties on 'iterative methods used for scheduling objects, especially those in which iteration counter is used as a variable in the computation' (Goldberg and Kay, 1977).

3.2.3 Recursion

Recursive functions are defined in terms of themselves. The well-known factorial function for example is defined recursively by:

```
fact (0) = 1
fact (n) = fact (n - 1)*n
```

Whereas iterative programming is based on the descriptions of actions modifying the situation, recursive programming describes the relationships between successive states, without expressing the computation strategy. The most important functionality of recursive procedures intervenes in program decomposition, i.e. in the representation of the problem domain as a part-whole relationship. This decomposition calls for two kinds of conceptual constructs: autoreference (declarative aspect) and nesting (procedural aspect) (Rouchier, 1987).

Recursive programming is very hard to learn and to teach. Studies have consistently shown that even on very simple problems students have enormous difficulty learning recursion. Beginners tend to use an iterative model of recursion (Kurland and Pea, 1983; Kessler and Anderson, 1986; Pirolli, 1986; Rouchier, 1987; Wiedenbeck, 1989). This type of model is compatible with tail recursion (the recursive calls are not nested when executed) but fails for full recursion (the recursive calls are nested when executed). The reasons for these difficulties are two-fold. First, there is no natural everyday activity that can serve as a precursor for recursion. Second, the dynamic model of iteration is (overly) salient and constitutes an obstacle for students in moving to this new program schema (Kessler and Anderson, 1986). A dynamic model may at times be useful for program design (Pirolli, 1986). The learning mechanisms such as 'by doing' or 'by example' are not sufficient for thorough understanding of recursion. This kind of situation reinforces novices' existing dynamic models.

Data is scarce or virtually non-existent on how expert programmers design and understand complex recursive programs. However, a crucial factor in programming recursion appears to be the use of static representations of situations in a functional or relational point of view of programming. It seems difficult to start a training process with novices directly with this type of representation. There is some empirical evidence, however, suggesting that more appropriate teaching situations can be designed. In particular these should not employ an iterative model but rather should start with full recursion problems (in contrast to the Kessler and Anderson (1986) suggestion to use iteration as a source of analogy. These would make the relationship between the structure of the programs and the structure of the problem explicit (for instance in a Logo procedure the places of recursive calls in the text of the program are related to the order in which the objects are produced). In these situations the emphasis is on the static properties of the procedure. In this case students can succeed in constructing more-effective strategies in program design (Rouchier, 1987).

3.3 Variables, data structures and data representation

Variables are the (proverbial) tip of the iceberg of the various entities that can be defined and managed in the programming process. Whatever the programming language, programmers must deal with the questions of available primitive entities (objects and operations on them), and how such entities should be modified to represent the elements of the domain problem. A key factor in the acquisition of programming knowledge is mastery of the relationship between the functional role of entities and their successive values during program running. This involves the acquisition of notions organized in a hierarchical fashion with strong connections between the notions of functions and variables (or predicates and variables in logical languages such as Prolog, objects and messages in object-oriented languages).

Numerous errors found in novice control structure protocols can be accounted for by the conception novices have of variables. Defining a variable as a name (related to the role played in the domain problem) or as an address (related to the role played in the program) is the first step towards understanding (Samurçay, 1989). When the value of a variable is modified, its naming and the functional relationship with other program elements remain invariant. At this level some of the errors made by novices involve the use of local rather than functional properties.

Variables also differ with regard to their functional meaning in student planning. 'External' variables correspond to the values controlled by the program user, i.e. they are explicit inputs (data) or outputs (results) of the domain problem. 'Internal' variables are controlled by the programmer, and produce intermediary results. Internal variables may be irrelevant for 'hand solutions', which is a source of error in novice programming (Samurçay, 1989). When a variable plays more than one role in the problem, errors are observed on the program fragment it appears in (Spohrer *et al.*, 1985).

A higher-level concept is required when variables occur in iterative or recursive programs in imperative languages. In this case, the variable is no longer an address (with a value) but needs to be seen as a function of execution, or a sequence of values. At this stage, the algebraic model of variable and equality novices may transfer from physics or mathematics fails as a mental model. A similar interaction has been observed in Prolog between 'logical' definitions and effective values of variables and functions. Novices encounter difficulties in understanding that the specific function defined in a given clause is determined by the instantiation of the inputs (Taylor and du Boulay, 1987).

This feature is closely related to difficulties teachers frequently notice when trying to clarify the difference between local and global variables, although one of the requirements for computer literacy is the learning of the use of procedures and functions to write modular programs. What is masked behind the distinction between local and global variables is the notion of context. Its acquisition calls for a thorough understanding of the 'notional machine'. Understanding backtracking as well as full recursion are related to the acquisition of this complex notion.

Dupuis and Guin (1989) have analysed how students use coding variables in a complex programming task that introduces recursive procedures in Logo. Two modes of coding were observed: a 'descriptive' code (different objects correspond to independent different variables) and an 'analytic' one (expressions with a single variable can be used to represent various related objects). The choice of mode is dependent on the mental model students generate of the computer device. In the descriptive mode coding variables are managed by the programmer, and are linked to an executive model. In the analytic mode, computing the values of variables is under computer control. This mode is related to a more static model of programming and is the only mode compatible with recursivity.

3.4 Programming methods

A method can be defined as an explicit aid for strategy research and management in solving problems of a given class. Implementing a method is dependent upon the relationship between type of problem and programming paradigm, and on the programmer's knowledge. Methods can provide help on various subtasks in program design (from the initial problem analysis to program documentation) and at various levels of problem decomposition (see Chapter 4.2). Up to now programming methods have only been used in training professional programmers, although there are needs for programming methods even in early acquisition. Van Someren has stressed this point for languages like Prolog which are not process oriented (van Someren, 1985). Existing methods share a number of points. All prescribe a top-down strategy which necessitates working on high-level before low-level structures. In terms of problem solving they are either prospective (analysis oriented by input data) or retrospective (analysis oriented by results). A recent overview of programming methods (Rogalski *et al.*, 1988) indicates that this particular area is characterized by its disproportionately low empirical data with respect to the high number of unanswered questions.

Studies on the use of methods by beginner programmers (Hoc, 1983; Morais and Visser, 1987) show that 'top-down' structured methods are difficult to use because beginners' spontaneous strategies are based on mental execution (from the input data). Rist (1986) reports that in the early stages of learning programming, the plan structures used by novices are action oriented (rather than object oriented as in expert plan structures). Even programmers familiarized with structured programming resort to decomposition oriented by processing and linked to mental execution in cases of increased problem difficulty (Hoc, 1981; Ratcliff and Siddigi, 1985). Professional programmers may also experience difficulty in using aids for program design because they lack sufficiently rich and flexible data structures to handle the requirements of a retrospective strategy (Hoc, 1988b).

These two points suggest that: (1) the representation and processing dimensions involved in the programming task (Figure 2) both enter into the issues surrounding teaching and learning methods; (2) an overly dominant dynamic model of programming can be an obstacle to the use of existing aids including models. Introducing models early on in training may help prevent negative reinforcement of the spontaneous computational programming paradigm.

Data from fields other than programming (mathematics and emergency management) indicate that methods can be taught to novice students in certain conditions. With respect to programming, the most critical condition is the existence of sufficient prior knowledge in the student (computer literacy) and use of teaching of situations containing relatively complex problems which call for making choices in data structuring, task organization and task distribution among students. This might involve a transposition of certain properties of 'programming at large' which would encompass the programming task as a whole and would lend meaning to programming methods and tools.

4 Conclusion

Acquiring and developing knowledge about programming is a highly complex process. It involves a variety of cognitive activities, and mental representations related to program design, program understanding, modifying, debugging (and documenting). Even at the level of computer literacy, it requires construction of conceptual knowledge, and the structuring of basic operations (such as loops, conditional statements, etc.) into schemas and plans. It requires developing strategies flexible enough to derive benefits from programming aids (programming environment, programming methods).

Taking the effective device into account calls for the construction of new systems of representation and processing in order to conceptualize the properties of the 'notional machine' and lead to appropriate human-computer interaction in programming. The familiar RPS related to action execution is a precursor that plays both a positive 'productive' role and a negative 'reductionist' one, assigning meaning to initial programming notions and becoming an obstacle when more static representations are needed.

Studies in the field and pedagogical work both indicate that the processing dimension involved in programming acquisition is mastered best. The representation dimension related to data structuring and problem modelling is the 'poor relation' of programming tasks. This reflects the current emphasis on the computational programming paradigm, linked to dynamic mental models, where programs are considered to be dynamic data processing, and procedures and functions are analysed as operations modifying data instead of defining new entities.

More research is needed on the acquisition of adequate 'static' representations, clearly required for the acquisition of functional or relational languages, but also necessary for efficient use of the powerful recursive tool. The recent stress on the necessity of some 'unlearning' by professional programmers testifies to the fact that focusing on dynamic aspects of programming may constitute an obstacle to further acquisitions. The major challenge today lies in the design of new paradigms in casual and professional programmers' training that will enable them to become more aware of the existence of various programming paradigms, more familiar with them, and more flexible in their representations, in order to open up the field of possible programming choices.

References

Anderson, J.R., Farell, R. and Sauers, R. (1984). Learning to program in LISP. *Cognitive Science,* **8**, 87-129.

Bonar, J. (1984). *Bridge: an intelligent programming tutor/assistant.* Technical Report. LRDC. University of Pittsburgh.

Bonar, J. and Soloway, E. (1985). Preprogramming knowledge: a major source of misconceptions in novice programmers. *Human-Computer Interaction,* **1**, 133-161.

Dupuis, C. and Guin, D. (1989). Gestion des relations entre variables dans un environnement de programmation LOGO. *Educational Studies in Mathematics,* **20**, 293-316.

Gilmore, D. and Green, T.R.G. (1984). Comprehension and recall of miniature programs. *International Journal of Man-Machine Studies,* **21**, 31-48.

Goldberg, A. and Kay, A. (1977). Methods for teaching the programming language Smalltalk. Technical Report, SSL-77-2. Xerox Palo Alto Research Center.

Green, T.G.R. (1980). Ifs and Thens: is nesting just for birds? *Software Practice and Experience,* **10**, 371-381.

Hoc, J.-M. (1981). Planning and direction of problem-solving in structured programming: a comparison between two methods. *International Journal of Man-Machine Studies,* **15**, 363-383.

Hoc, J.-M. (1983). Analysis of beginner's problem-solving strategies in programming. *In* T.R.G. Green, S.J. Payne and D. van der Veer (Eds), *The Psychology of Computer Use.* London: Academic Press, pp. 143-158.

Hoc, J.-M. (1988a). *Cognitive Psychology of Planning.* London: Academic Press.

Hoc, J.-M. (1988b). L'aide aux activités de planification dans la conception des programmes. *Le Travail Humain,* **51**, 323-333.

Hook, K., Taylor, J. and du Boulay, B. (1988). Redo "try once pass": the influence of complexity and graphical notation on novices' understanding of Prolog. Cognitive Science Research Reports No. 112. Brighton: University of Sussex.

Johnson, L.W., Soloway, E. (1983). Proust: Knowledge-based programming understanding. Technical Report. New Haven: Yale University.

Katz, I.R. and Anderson, J.R. (1988). Debugging: an analysis of bug-location strategies. *Human-Computer Interaction,* **3**, 351-399.

Kessler, M.K. and Anderson, J.R. (1986). Learning flow control: recursive and iterative procedure. *Human-Computer Interaction,* **2**, 135-166.

Kurland, D.M. and Pea, R. (1983). Children's mental models of recursive Logo programs. *Proceedings of the 5th Annual Conference of the Cognitive Science Society.* Rochester, NY: University of Rochester.

Laborde, C., Balacheff, N. and Mejias, B. (1985). Genèse du concept d'itération: une approche expérimentale. *Enfance,* **2-3**, 223-239.

Morais, A. and Visser, W. (1987). Programmation d'automates industriels: adaptation par les débutants d'une méthode de spécification de procédures automatisées. *Psychologie Française,* **32**, 253-259.

Ormerod, T.C., Manketelow K.I., Steward, A.P. and Robson E.H. (1988). The effects of content and representation on the transfer of PROLOG reasoning skills. *Proceedings of the International Conference on Thinking.* Aberdeen.

Pea, R.D. (1986). Language independent conceptual "bugs" in novice programmers. *Journal of Educational Computing Research,* **2**, 25-36.

Pirolli, P. (1986). A cognitive model and computer tutor for programming recursion. *Human-Computer Interaction,* **2**, 319-355.

Ratcliff, B. and Siddigi, J.I.A. (1985). An empirical investigation into problem decomposition strategies in program design. *International Journal of Man-Machine Studies,* **22**, 77-90.

Rist, R.S. (1986). Plans in programming: definition, demonstration and development. *In* E. Soloway and S. Iyengar (Eds), *Empirical Studies of Programmers.* Norwood, NJ: Ablex.

Rogalski, J. and Hé, Y. (1989). Logic abilities and mental representations of the informatical device in acquisition of conditional structures by 15-16 year old students. *European Journal of Psychology of Education,* **4**, 71-82.

Rogalski, J. and Samurçay, R. (1986). Les problèmes cognitifs rencontrés par des élèves de l'enseignement secondaire dans l'apprentissage de l'informatique. *European Journal of Psychology of Education,* **1**, 97-110.

Rogalski, J. and Vergnaud, G. (1987). Didactique de l'informatique et acquisitions cognitives en programmation. *Psychologie Française*, **32**, 275-280.

Rogalski, J., Samurçay, R. and Hoc J.M. (1988). L'apprentissage des méthodes de programmation comme méthodes de résolution de problème. *Le travail Humain*, **51**, 309-320.

Rosson, M.B. and Alpert, S.R. (1988). The cognitive consequences of object-oriented design. IBM Research Report RC 14191.

Rouchier, A. (1987). L'écriture et l'interprétation de procédures récursives en Logo. *Psychologie Française*, **32**, 281-285.

Samurçay, R. (1985). Learning programming: an analysis of looping strategies used by beginning students. *For the Learning of Mathematics*, **5**, 37-43.

Samurçay, R. (1986). Understanding the cognitive difficulties of novice programmers: a didactical approach. *3rd. European Conference on Cognitive Ergonomics.* Paris.

Samurçay, R. (1987). Plans et schémas de programmes. *Psychologie Française*, **32**, 261-266.

Samurçay, R. (1989). The concept of variable in programming: its meaning and use in problem solving by novice programmers. *In* E. Soloway and J.C. Spohrer (Eds), *Studying the Novice Programmer.* NJ: Lawrence Erlbaum.

Samurçay, R. and Rouchier, A. (1985). De "faire" à "faire-faire": la planification de l'action dans une situation de programmation. *Enfance*, **2-3**, 241-254.

Sime, M., Arblaster, A. and Green, T.R.G. (1977). Reducing programming errors in nested conditionals by prescribing a writing procedure. *International Journal of Man-Machine Studies*, **9**, 119-126.

Soloway, E. and Ehrlich, K. (1984). Empirical studies of programming knowledge. *IEEE Transactions on Software Engineering*, **5**, 595-609.

Soloway, E., Ehrlich, K., Bonar J. and Greenspan J. (1982). What do novices know about programming? *In* B. Schneiderman and A.Badre (Eds), *Directions in Human-Computer Interaction.* Norwood, NJ: Ablex.

Spohrer, J.C. and Soloway, E. (1986). Analyzing the high frequency bugs in novice programs. *In* E. Soloway and S. Iyengar (Eds), *Empirical Studies of Programmers.* Norwood, NJ: Ablex.

Spohrer, J.C., Soloway, E. and Pope, E. (1985). A goal-plan analysis of buggy Pascal programs. *Human-Computer Interaction*, **1**, 163-207.

Taylor, J. and du Boulay, B. (1986). Studying novice programmers: why they may find learning Prolog hard. Cognitive Science Research Reports No. 60. Brighton: University of Sussex.

Taylor, J. and du Boulay, B. (1987). Learning and using Prolog: an empirical investigation. Cognitive Science Research Reports No. 90. Brighton: University of Sussex.

van der Veer, G. and van der Wolde, J. (1983). Individual differences and aspects of control flow notations. *In* T.R.G. Green, S.J. Payne and D. van der Veer (Eds), *The Psychology of Computer Use.* London: Academic Press.

van der Veer, G., van Beek, J. and Gruts, G.A.N. (1986). Learning structured diagrams. Effects of mathematical background, instruction and problem semantics. *Visual Aids in Programming,* Passau.

van Someren, M. (1985). Beginners problems in learning PROLOG. Technical Report. Department of Social Science Informatics, University of Amsterdam.

Widowski, D. and Eyfert, K. (1986). Representation of computer programs in memory: comprehending and recalling computer programs of different structural and semantic complexity by experts and novices. *3rd European Conference on Cognitive Ergonomics.* Paris, pp. 93-103.

Wiedenbeck, S. (1989). Learning iteration and recursion from examples. *International Journal of Man-Machine Studies,* **30**, 1-22.

Chapter 2.5

Programming Languages in Education: The Search for an Easy Start

Patrick Mendelsohn[1], T.R.G. Green[2] and Paul Brna[3]

[1] *Faculty of Psychology and Educational Science University of Geneva, Switzerland*
[2] *MRC Applied Psychology Unit, Cambridge, UK*
[3] *Department of Artificial Intelligence, University of Edinburgh, UK*

Abstract

(i) We first discuss educational objectives in teaching programming, using Logo research as a vehicle to report on versions of the 'transfer of competence' hypothesis. This hypothesis has received limited support in a detailed sense, but not in its original more grandiose conception of programming as a 'mental gymnasium'. (ii) Difficulties in learning quickly abnegate educational objectives, so we next turn to Prolog, which originally promised to be easy to learn since it reduces the amount of program control that the programmer needs to define, but which turned out to be very prone to serious misconceptions. Recent work suggests that Prolog difficulties may be caused by an inability to see the program working. (iii) Does the remedy therefore lie in starting learners on programmable devices that are low level, concrete and highly visible? Research on this line has brought out another problem: learners find the 'programming plans' hard to master. (iv) Finally, we sketch a project designed to

Psychology of Programming
ISBN 0-12-350772-3

teach standard procedural programming via 'natural plans'. Our conclusions stress pragmatic approaches with much attention to ease of use, avoiding taking 'economy' and 'elegance' as virtues in their own right.

1 Introduction

For all the efforts of the computing fraternity, computing remains inaccessible to many people, especially the old, the very young, the disabled, and those with low educational achievements. In this chapter we shall outline some of the attempts to open up computing for educational purposes, especially school children and 'distant learners' – pupils learning without face-to-face contact with a tutor. For these purposes one wants to avoid languages that enforce many new abstract ideas, or have too many 'programming tricks' to be learnt, or give too little immediate concrete feedback. So there has been much interest in new languages designed for children that are friendlier and more concrete, such as Logo; and in languages like Prolog, which appear to avoid some of the problems of abstractions and programming tricks. We shall describe work with both of these languages. We shall also look at attempts to break out of the 'programming language' mould and to create systems that are more immediate than either Logo or Prolog.

At the same time as new languages have been developed, there has been much discussion of whether the primary aim should be to teach children programming for its own sake, or to use programming in the service of some other end or discipline – 'programming to learn, or learning to program'. Different programming cultures have emphasized different balances between the two polar positions. These educational objectives are not, strictly speaking, within the scope of this volume, which takes as its starting point the view that unless the programming language is adequately matched to the abilities of its users, nothing else can be done and all objectives will fail. Yet, as we shall quickly observe, these aspects tend to get themselves mixed together.

Of course, choosing a language in school is not a free choice. The development of educational computing is directly dependent on different sources of material constraints, both institutional and human. Teachers who decide to use a computer in the classroom have, at least initially, only a limited influence on these constraints. They passively follow developments in hardware and software and adapt to political choices concerning computer equipment more than participating in decisions. They accompany human transformations more than determining them. Nevertheless, there are enough pedagogical alternatives remaining for any reflection on the adaptation of computer languages to the goals of teaching to be useful and fruitful.

The rapid and spectacular progress made in computer and software performance suggests that great care should be taken when deciding on which equipment to introduce into the classroom for teaching programming. How and with what goals in mind must programming languages be used at school such that this utilization and teaching are not completely outdated in five years time? It quickly becomes clear that even if organizational constraints can be put aside, language choices must be based on many different criteria.

First, the *language* level will affect the level at which the pupils will work, and will partly determine how much of their time is spent on learning to program and how much on using that knowledge to aid other kinds of learning.

Secondly, there is the *transfer* problem of making skills and knowledge learned in the context of programming available in a different context. Kurland *et al.* (1989) point out that the degree of transfer depends on how close the domains are. 'Near' transfer effects can be obtained without too much difficulty (Littlefield *et al.*, 1988), but 'far' transfer is difficult to establish unless the analogy between domains of knowledge is explicitly taught.

Finally, the teacher must choose a *teaching style*. The first applications of computing in education were marked by a debate, often lively (Solomon, 1986), between the defenders of programmed teaching and the partisans of learning through discovery and self-teaching. For the first clan, computing is essentially an effective tool for training and repeating teaching sequences (Suppes, 1979). For the others, computing is more like a medium, a support for elaborating environments in which the child is his or her own knowledge builder (Papert, 1980).

1.1 Language cultures and research questions

Given these complex interrelationships, it is hardly surprising that the choice of a classroom language brings with it a particular culture. 'If you weren't interested in the problems that Logo can deal with, you wouldn't have chosen Logo' – and thus, further research on Logo tends to be dominated by the same set of questions. In such a way does a culture perpetuate itself. And more: the Logo culture has always emphasized interaction, either with real devices ('turtles') or else with virtual devices ('screen turtles'); these crawling and drawing devices add amusement but also present powerful challenges. Logo systems without graphics are unthinkable. Not so Prolog, which has always emphasized reasoning about a database of assertions; many Prolog systems have no graphics, or limited graphics that tend to subvert the declarative style of Prolog with a procedural outlook.

Problems with the Logo language itself have not been a major discussion point. There are some predictable difficulties, such as learning to tell the turtle which way to go when it is facing south (when the turtle's left is the viewer's right), but, in general, turtle graphics has been fairly successful. It is not surprising, then, that the major research efforts in connection with Logo have dwelt on 'the cognitive effects of learning to program', in Pea and Kurland's phrase.

Prolog, on the other hand, has been a troublesome language, and fewer interesting research results have come out of teaching children Prolog. The reasons appear to be an emphasis on relational database querying, which is relatively unproblematic; emphasis on teaching some other subject using Prolog simply as a means to represent the knowledge; and the use of a 'toolkit' as a front end to give a simplified way in which children can add knowledge to a ready-programmed computational mechanism. These can all be seen as ways to get some benefits from Prolog while not having to teach children how to program in Prolog.

Our first two sections, therefore, will deal with research in each of these traditions. We then present some contrasting work in which the traditional, extremely abstract concept of a programming language has been rejected in favour of far more concrete devices, in one case a microprocessor with sensors and in the other, a model train. Certain themes will recur throughout these treatments, such as the problem of helping novice programmers to perceive programming plans, and we finish with a forward-looking research project which aims to present plans more directly than has been achieved before.

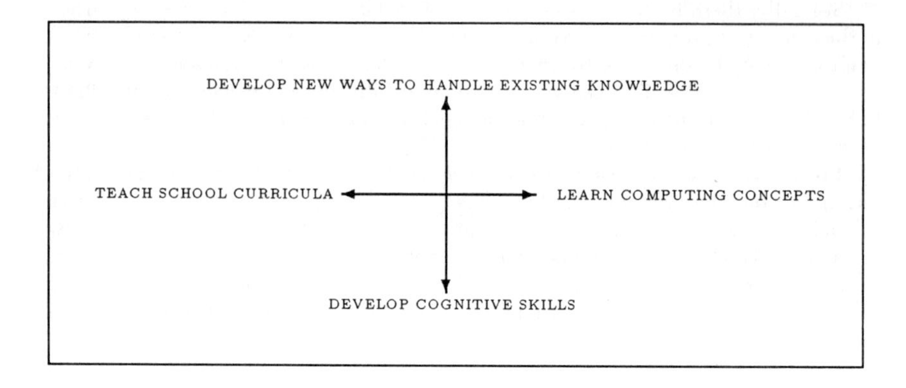

Figure 1: Schematic representation of different objectives for programming activities in school: vertical axis, transfer of competence hypothesis; horizontal axis, the learning of content.

2 Programming to learn, or learning to program? Educational objectives and Logo

Logo is the language that is predominantly used in general education in France and is also widely used in Britain and North America. It has immediate appeal to children because of its 'turtle graphics', which originally were instructions driving a small device that crawled around the floor. The device could lower a pen to mark its path and had sensors to record obstructions. Interest in the possibilities of Logo's graphics model gradually became one of the dominant themes of Logo work, and the floor-based turtle is now less important than the screen turtle which draws lines on the VDU. The same instructions are used, preserving the 'turtlocentric' view: the turtle knows how to turn left or right by a stated number of degrees, and how to go forwards or backwards by a given length. Logo is a procedural language with provision for procedures (using parameters called by value), variables, conditionals, recursion and graphics. Data structures include numbers, strings and lists.

Following the guiding principles described in our introduction we shall devote our attention to the central question raised by this choice of language, namely, the problem of deciding upon educational objectives. Four types of project will be presented; we believe that the reflections proposed here can readily be generalized to many different programming contexts.

The four approaches we shall describe can be represented on two axes (Figure 1). The options that consist of using Logo, with the 'transfer of competence' hypothesis as a guiding concept, are represented on the vertical axis. The research scientist is more interested here in programming as the exercising of competence that can be reinvested in other school situations, than programming as an activity of developing programs.

Represented on the horizontal axis, on the other hand, are the practices of teachers who use programming languages with the central idea that children learn, above all, specific information and know-how related to programming.

Before going further into the details of this analysis we can already note that what actually does happen in the classroom corresponds more to the horizontal axis, whereas the vertical axis (transfer hypothesis) is generally called upon to justify these practices!

2.1 The transfer of competence hypothesis

The general hypothesis of transfer covers, in fact, two quite different conceptions. On the one hand there is the most classical position at the origin of the creation of Logo. A programming language is considered as a medium that creates new ways of dealing with existing knowledge. This perspective is supported by several authors (Papert, 1980; O'Shea and Self, 1983; Solomon, 1986; Lawler, 1985). These authors stress the role of self-training and the child's discovery of his or her own problem-solving strategies as well as in co-operative problem solving by learners exploring a programming world that is shared between work stations. They also consider the introduction of programming to be a revolution in teaching practice. Knowledge is transformed and the classical competence taught at school is generally rendered obsolete. Problems of evaluation are generally circumvented in that it is considered that if one child can do something, other children can do the same if they are given the means.

On the other hand there are those authors who support a more cognitivist perspective. Through programming practice children develop cognitive skills that are identifiable and transferable to other situations. The most studied of these elementary aptitudes solicited by programming activity have been analogical and temporal reasoning, mathematical operations (Ross and Howe, 1981), the planning of action (Pea and Kurland, 1984; Lehrer *et al.*, 1988), error correction (Klahr and Carver, 1988), and the development of logical and spatial operations (Mendelsohn, 1986).

2.2 Transfer of competence as new approaches to knowledge and learning

Within Papert's perspective, Logo provides an easy approach to the art of heuristics. The principal force of structured programming is that procedures can be created, like separable blocks, to obtain a progressive construction of the solution or for solving larger problems (an example of this is provided in the classical procedure HOUSE given to beginners). It is therefore logically possible to know how each one of us breaks down a problem into simple units and then co-ordinates these units into macro-actions.

Howe and O'Shea (1978) provide an illustration of this approach. These authors attempted to test the hypothesis according to which the child learning to program in Logo uses a system of strong metaphors to describe reality. These metaphors can be linked to the body schema (using Logo to add meaning to and describe movements), to denomination (the fact that procedures have a name and can be recalled in another part of the program), to the breaking down of a problem into subproblems (structured programming). Why not therefore conclude that the child can then use these metaphors to transmit knowledge to someone else in a different context?

Howe and O'Shea designed the following experiment to test this hypothesis. One group of children having learnt to program in Logo and another group with no programming experience were placed in a situation similar to the game 'battleships' in front of a screen hiding a partner. The subjects have in front of them a figure composed of small geometrical shapes like in a jigsaw puzzle and their partners have a set of geometrical shapes, a subset of which is identical to those used in the figure. The subjects have to explain to their partners, without showing them any of the shapes, how to build the figure.

Howe and O'Shea hypothesized that if the children with programming experience in Logo have learnt anything relative to the metaphors of communication (modularization, denomination, sequentiality, etc.), then they should be more capable of explaining how to build the figure than the children with no programming experience. The authors expected the 'programming' children to break down the figure into relevant subsets, give names to these subsets using suitable labels and only to propose executable actions to their partner. The 'non-programming' children were expected to enumerate the different pieces of the puzzle giving ambiguous instructions.

The results show that some children do seem to have recognized the analogy between this problem and the programming situation. The results must be considered with caution, however, as the necessary methodological precautions were not taken to allow a non-subjective interpretation of the data. Other similar experiments have not shown such analogical transfer (Littlefield *et al.*, 1988). Nevertheless, these considerations and the pedagogical reflections associated with these experiments have a non-negligible impact on teachers and can encourage them to re-evaluate some of the pedagogical objectives of their classes.

2.3 Transfer of competence as the development of cognitive skills

The other perspective of the 'transfer hypothesis axis' is characterized by a more rigorous experimental approach and by its explicit reference to cognitive psychology. A 'cognitive skill' is a competence associated with the manipulation of identifiable operators, not specific to programming. Moreover, this competence must be sufficiently generalizable and exercised to be re-invested in other tasks. Pea and Kurland's (1984) study on the development of aptitude for action planning is the most representative of this type of research. This aptitude is evaluated by the capacity for optimizing the representation of a string of actions that is too large to be managed directly in working memory. These authors have experimented with seven- and eleven-year-old children some of whom had one year of Logo instruction at the rate of one hour a week. For Pea and Kurland, planning is only necessary if the situation imposes several constraints on the subject: (1) planning is the only means of solving the problem; (2) the task must be sufficiently complex so that memorizing the subgoals is impossible; and (3) the area of knowledge is familiar enough for the children to be able to identify the elementary actions to be performed. The situation-problem used by these authors was represented by a three-dimensional model of a classroom with objects and furniture (chairs, tables, plants, etc.). The starting point was the door of the classroom. The children had to carry out a number of concrete actions on the model: water the plants, clean the blackboard, put the chairs in front of the tables, wash the tables, move certain objects, etc. These different actions can be performed in any order but the children were instructed to find the route that minimized the number of movements necessary. Each subject had three trials at the beginning of the school year (pre-test) and three more trials at the end of the year (post-test). A measure of performance was calculated for each subject. The results provided little support for the transfer hypothesis. The expected effects of age and order of trial were observed but no significant difference was observed between children with Logo experience or not (neither in terms of the types of strategy used nor the objective measures of efficiency).

In a second experiment Pea and Kurland provided a further analysis of this problem. As the transfer task in the first experiment may have been too different from the type of task performed in programming situations, a computerized version of this task was used in the second experiment. Thus, instead of being performed on a scale model, the operations were simulated on a computer and the subjects had to provide a list of commands. Even if this situation penalizes the control group the results were no better than in the previous experiment. Similar results were obtained by Littlefield *et al.* (1988) using the same planning situation.

These results, disappointing for those who defend the transfer of competence hypothesis, lead to the formulation of at least three statements (for a detailed analysis see Crahay, 1987; Mendelsohn, 1988; Kurland *et al.*, 1989).

* It is unrealistic to think that one hour a week of programming over a one-year period will allow transfer of such a specialized competence as planning. Would anyone have attempted a similar experiment for a less attractive activity, such as the game of chess, where no programming language is involved?

* One should perhaps be more realistic and take more time analysing what children really do before hypothesizing about the expected form of transfer. Pea and Kurland (1984) also stress the methodological problems associated with inaccurate evaluation of the level of expertise that the 'programming' children attain. To surmount this difficulty Klahr and Carver (1988) suggest that a formal analysis be made of what is learnt and what is supposed to be transferable. However, this costly precaution is rarely taken into considerations by the researchers in this field.

* The principal interest of Logo in the classroom is perhaps not to be found in this direction. A programming language is above all a system of representation and the interest of such a system lies in its capacity to highlight new properties of the manipulated objects while allowing the automatic execution of complex processing. Learning arithmetic benefits the child more by proposing a formal language that facilitates the symbolic processing of complex numerical problems than by developing hypothetical aptitudes for reasoning. The same is perhaps true for learning to program. This point of view characterizes the second axis of Figure 1.

2.4 The acquisition of new knowledge

With respect to the axis that represents the learning of content hypothesis (the horizontal axis of Figure 1), it is also possible to oppose two approaches with very different goals. On the one hand there is the idea that programming helps children learn, above all, concepts of computing: programming operations, data structures, variables (reference to the term learn). On the other hand, there are those who stress the teaching of specific courses (programming is always applied to particular content). Thus programming is used as a back-up for teaching geometry, arithmetic, or even grammar.

2.5 Knowledge acquisition as mastering computer concepts

The teaching of programming at school is not aimed at producing computer programmers but rather computer users. One can nevertheless think that basic computer concepts are an integral part of computer culture. Because of this, the learning of these concepts has received much attention from psychologists who have considered this to be a particularly rich field for fundamental research. The development of these concepts in children, the cognitive difficulties involved in their acquisition, and their origin, are the main themes of research that interest computer educationalists (Rogalski and Vergnaud, 1987). Some examples of the operations studied here are iteration (Soloway *et al.*, 1983; Kessler and Anderson, 1986), conditional branching (Rogalski, 1987) and recursion (Anzai and Uesato, 1982; Rouchier, 1987; Mendelsohn, 1985; Pirolli, 1986; Kahney, 1983). Some more-simple concepts can be added to this list: sequentiality, modularity, and the notion of computer variable. These have been studied less since they are implicitly linked to the former.

Recursion interests many researchers because of its particular status in programming. It is a powerful operation which is trivial in its definition yet raises many problems when it comes to teaching it in its most complex forms. Hofstadter (1979) describes recursion as a mode of reasoning characterized by self-reference and the nesting of processing ('This sentence has five words'). Roberts (1986) stresses the algorithmic point of view. Recursion is thus a means of solving a problem by reducing it into one or several subproblems which are (1) structurally identical to the original problem, and (2) more simple to solve. Finally, strictly from a computing point of view, recursion is a program control structure (its written form depending on the language used). Kurland *et al.* (1989) stress the fact that it is important for the learner to have a good representation of what is going on inside the 'black box' when running recursive procedures in order to be able to make good use of them.

With a similar aim, we wanted to combat children's difficulties in understanding recursive procedures: we developed (Mendelsohn, 1985; Rouchier, 1987) a teaching technique that involves recognizing the specific figural characteristics that allow a graph to be described by recursion. The model of reference chosen for this is central recursion, which can be described formally as follows:

```
TO PROCEDURE-NAME :VAR
IF predicate [ACTION3 STOP]
ACTION1
PROCEDURE-NAME :VAR
ACTION2
END
```

(with `ACTION2` non-null in order to rule out tail recursion).
The prototypical example used to help children understand the functioning of this recursive procedure is the example of the ski tow. It is possible to imagine a series of ski tows that take the skier to the top of the slopes. Each time the skier takes a ski tow (ACTION1) a ski down (ACTION2) is potentially accumulated, the parameters of which are related to ACTION1 (difference in height, for example). Performing ACTION2 is suspended and only carried out once the skier decides to stop climbing. At the top of the slopes the skier can perform any other kind of action such as rest or eat (ACTION3) and then ski down the slopes in the opposite order in which they

were climbed. Using this model one can vary during learning both the existence, content and structure of the procedures ACTION x, and the number and nature of the transformation functions associated with the variables. The teaching is therefore centred on the conceptualizing of regularities observed during these variations (Figure 2).

In this way the child learns to use recursion as a means of describing objects with very precise characteristics. In the examples provided one can underline the initial symmetry and the order in which the figure is constructed. The teacher can progressively introduce sources of variation that slowly add to the diagram until it is formally complete. This teaching technique is the same as used with other operations such as addition and multiplication in arithmetic. Teaching of computer programming can be seen in this way as a laboratory that allows testing relative effects of various instruction methods on student's mental models. This topic is now starting to be explored in the direction of guided discovery learning (Lehrer *et al.*, 1988).

2.6 Knowledge acquisition as teaching school curricula

This last theme deals with the use of Logo for teaching school curricula. It is the course content that is stressed here, the programming language being considered as a support for the representation of original properties concerning contents and transformations (Hoyles and Noss, 1987). For many teachers this is often the only project that incites them to use Logo in the classroom. Several original examples of this have been provided by Bourbion (1986). These are educational applications in which the control of the subject's performance is centred on the course being taught (mostly arithmetic and geometry). Programming thus becomes an implicit activity, mastery of which is often considered as an introduction to the main work.

The problem that has been chosen as an example here consists, first of all, in editing a small graphics program. With this program, two walls the same height can be built with bricks of different thickness in two places P1 and P2 (Figure 3, first phase). The algorithm places a brick at P1 then checks if the height of the wall is lower at P2. If this is not the case another brick must be added at P1 and the relative height checked again. If this is the case then a brick must be added to P2 and then a check made to see if the height at P1 is lower, and so on.

The second step (Figure 3, second phase) involves subtly transforming this program by removing all graphic aspects from the procedures. In this way we obtain an isomorphic program which by generalization becomes a procedure for calculating the lowest common denominator.

With this technique, arithmetic conceptualization re-adopts its principal aim, that is the progressive pruning of the actions that one is led to perform in reality in order to conserve only the essential transformations. The programming language thus acquires the status of a formal language and can even become, as such, an object of teaching (see also Vitale, 1987). Many other examples involving geometry concepts, physics or language could be mentioned. The recurrent idea remains that of using a programming language to represent specific components of the subject discipline and throw light on new links between different domains (mediated learning).

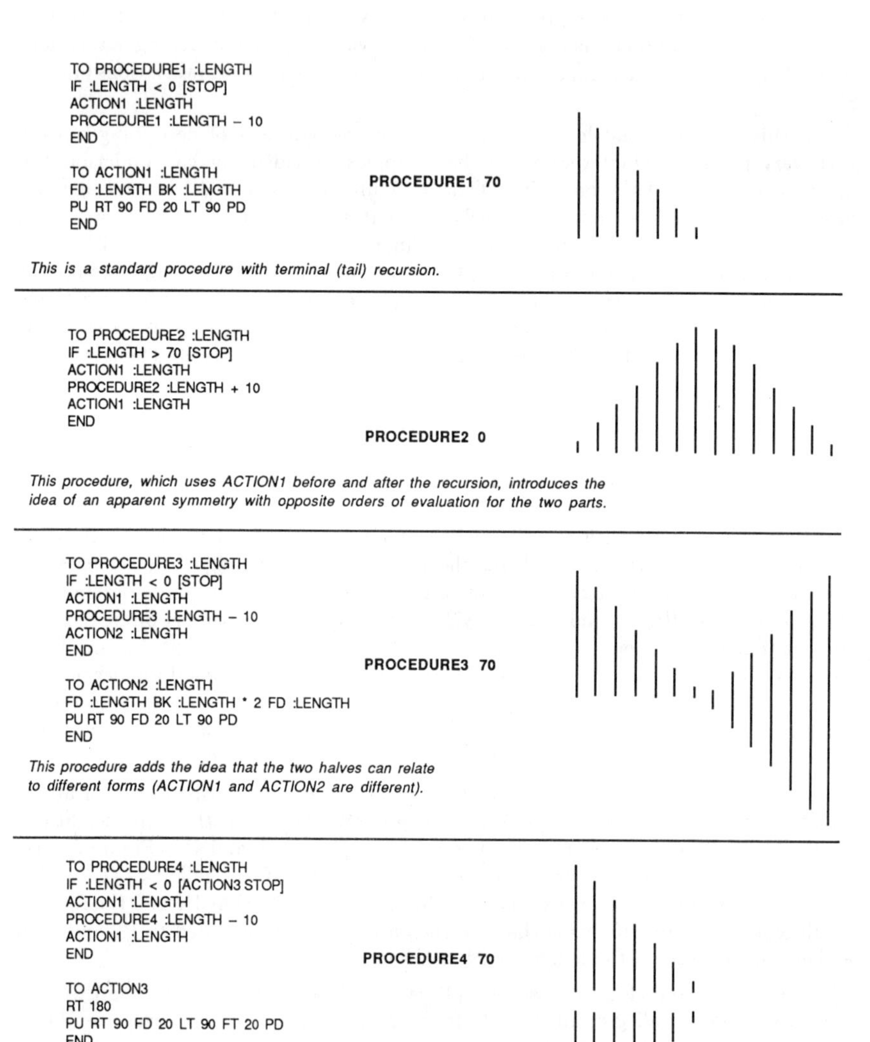

Figure 2: An example of progression based on a starting diagram for the writing of centrally recursive procedures in Logo.

First phase: Developing a procedure for building two walls of equal height from bricks of different thicknesses

```
TO CONSTRUCT
IFELSE :H1 < :H2 [WALL1] [WALL2]
IF :H1 = :H2 [STOP]
CONSTRUCT
END

TO WALL1                        TO WALL2
SETPOS [0 0]  FD :H1            SETPOS [0 0]  FD :H2
MAKE "THICKNESS :E1             MAKE "THICKNESS :E2
PLACE.BRICK                     PLACE.BRICK ................add 1 brick of given thickness (code not shown)
MAKE "H1 :H1 + :E1              MAKE "H2 :H2 + :E2
END                             END

TO WALLS :A :B
INIT .........................MAKE "H1 0    MAKE "H2 0    MAKE "E1 :A    MAKE "E2 :B
CONSTRUCT
RESULT....................PRINT SENTENCE "Height: :H1
END
```

Input

```
? WALLS 12 18
Height: 36
```

Output

Graphics output

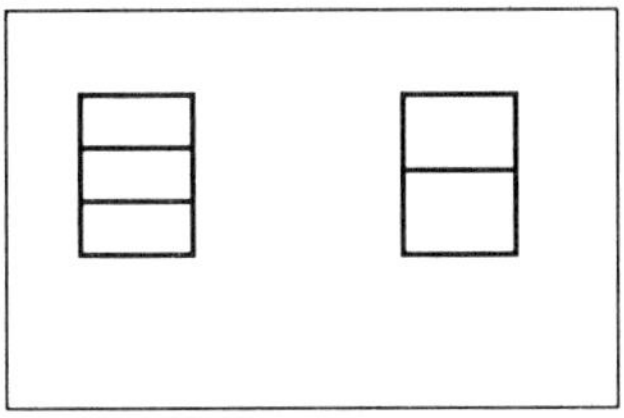

Second phase: The graphics instructions and semantic references are removed from the program, leaving only the calculations.

```
TO PART1                  TO PART2                  TO RESULT
MAKE "H1 :H1 + :E1        MAKE "H2 :H2 + :E2        PRINT :H1
END                       END                       END
```

The procedure for building two walls becomes a procedure for least common multiple:

```
TO LCM..................................(Lowest Common Multiple)
INIT
CONSTRUCT
RESULT
END
```

Figure 3: An example of teaching school curricula through Logo: 'Procedure for building two walls' (after Bourbion, 1986).

3 The misconception problem: Prolog

The unfortunate novice is prey to misconceptions when meeting any programming language. There have been excellent studies of misconceptions about even so 'small' a topic as assignment (Mayer, 1979; Putnam *et al.*, 1986; du Boulay, 1986). Intensive studies of misconceptions were made by the group working with Solo, a language designed for distance teaching of adult learners in the Open University (Eisenstadt, 1979): some of their work was put into analysing error messages and how these were misunderstood (Eisenstadt and Lewis, 1985), and other work was devoted to analysing faulty mental models of recursion and conditionality (Kahney, 1983).

All programming languages found in the classroom are prone to a wide range of misconceptions – but arguably, the language that has been found to engender the greatest number of problems is Prolog. The idea behind Prolog is that to build a program, the programmer just writes down true statements that capture some logical relations, and the Prolog interpreter can draw on these as needed to discover whether they imply the truth of some arbitrary proposition. The original shining hope was that there would be no need to understand what the Prolog interpreter did – or even that it did anything at all: the programmer would just write down the conditions that defined the required answer, and out it would pop.

For example, the logic of the `ancestor` relation can easily be described in words as 'somebody is your ancestor either (1) if that somebody is one of your parents, or (2) if that same person is the ancestor of one of your parents'. It can be written in Prolog as two distinct propositions, or *clauses*, as:

```
1.      ancestor(X,Y):- parent(X,Y).
2.      ancestor(X,Y):- parent(X,Z), ancestor(Z,Y).
```

where `ancestor(X,Y)` is read as 'Y is the ancestor of X', the symbol `:-` is read as 'if the following is true', and `parent(X,Y)` is read as 'Y is the parent of X'. The comma symbol has the logical reading of 'and the following is true' so that the second clause means 'there is a third person Z who is a parent of X, and Y is an ancestor of Z'.

This definition of `ancestor` would be used in conjunction with a database of facts about who is the parent of whom:

```
parent(abel, adam).
parent(abel, eve).
parent(cain, adam).
parent(cain, eve).
parent(enoch, cain).
...  etc...
```

Prolog is then able to deduce whether or not one person is the ancestor of the other. The programmer needs to know no 'programming tricks' about searching data bases: those are all built into the interpreter. By setting the *query*

```
?ancestor(enoch, eve)
```

Prolog will correctly state that 'eve' is an ancestor of 'enoch'.

Of course, the Prolog interpreter is not gifted with foresight or special knowledge: it simply performs conventional depth-first search over the database. Ideally, that should mean that the programmer need only worry about the 'logical' meaning of a program and not about the way that Prolog evaluates a program. This is to say that *the logical reading of a program is supposed to coincide with the behaviour of the Prolog system*. Unfortunately this is not the case – and the discrepancy between the logical reading and Prolog's behaviour in executing the program is one fairly well-known source of problems. (Example: if the subgoals of clause 2 are reversed, so that `ancestor` is tried before `parent`, the logical meaning may well appear unchanged – but Prolog will eventually enter an unending recursion.)

Until about 1980, the use of Prolog was restricted to a small number of research groups around the world. Prolog first became known to the primary and secondary educational communities through the work of Kowalski and Ennals (see Nichol *et al.*, 1988, for a recent account of this group's work). For the next few years, the advocates of Prolog made much of the possibilities inherent in a language in which it was fairly easy to describe logical relationships without having to worry quite so much about how to control computation. But, although the language is very powerful and can be used with great effect by some students, other students seem to become tremendously confused. One problem was that there were inconsistencies present in the teaching materials and some major conceptual problems with the language (Bundy, 1984). In fact, Pain and Bundy (1987) outlined over six different, partial accounts which were then in use!

Evidence relating to misconceptions associated with Prolog began to accumulate. For example, students had problems with *backtracking*, which is what Prolog does when a query fails. Students were also known to have problems with the way Prolog tries to match two data structures. This form of pattern matching is special to Prolog and is known as *unification*. For example, matching `ancestor(enoch, X)` with `ancestor(Y, eve)` would result in the variable `X` being bound to `eve` and the variable `Y` being bound to `enoch`. Coombs and Alty (1984) were quick to note these difficulties, and more-detailed work demonstrating particular misconceptions was reported by a variety of researchers (Ormerod *et al.*, 1986; Fung *et al.*, 1987; Taylor, 1987; van Someren, 1985).

In due course a lengthy taxonomy based on collected reports was produced by Fung *et al.* (1987). Many misconceptions were particular to Prolog and its control structure; others were more general, such as the 'meta-analysis' documented by many researchers – i.e. the belief that Prolog is able to 'foresee' which goals will and will not succeed, and can choose its behaviour accordingly. This was especially noted in the context of extremely familiar relationships, such as kinship relations (parent, cousin, aunt/uncle, etc.). Such relationships are frequently used in introducing Prolog – as we have done here, in fact – but their very familiarity can make trouble for some students.

We need to consider three questions: Where do misconceptions come from? Should they be avoided at all costs, or are they part of learning? Can they be avoided, or at least reduced?

3.1 Where do misconceptions come from?

Taylor (1987) has provided a detailed analysis of constructing a Prolog program in terms of a three-level discourse model, distinguishing between the general, logical and mechanistic levels. Each type of discourse has its own appropriate components to deal with inputting data, reasoning about the data, outputting the results, and doing 'reflective evaluation' (the process of reconsidering what one did and whether it could be different). This model allows her to categorize various pathways through the framework in terms of the discourse level adopted at each stage. She considers the activity of learning to program as beginning with the general problem-solving discourse level. The student then has to come to terms with the formal problem-solving discourse level. This formal level requires students to come to terms with both the level of logical discourse and also the level of mechanistic discourse. Not surprisingly, when an alternative discourse path offers itself, by performing reasoning at the 'real-world' level, students will be tempted to adopt it, leading themselves to ignore the actual behaviour of the program. One of the attractive features of Taylor's model is that it emphasizes that certain problems arise only because Prolog is so close to the general discourse level. If it were further removed, such 'meta-analytic' problems would arise less often.

Negative transfer from previous programming experience has been frequently put forward as an explanation for learners' difficulties. Van Someren (1988) found that many students seemed to have a simple algorithm for converting a Pascal-based program into a Prolog form. Their algorithm, which was faulty, appeared to account for several errors, Pascal assignment being a particular problem. White (1988) also followed this claim up in an attempt to find statistical evidence that some such effect exists. His results were strongly suggestive but fell short of statistical significance.

Yet a third possible origin for misconceptions is to be found in the unification process. Van Someren (1990) suggests that many of the errors made by students in connection with the unification process can be explained by a 'damaged' version of the correct algorithm together with a 'repair' mechanism for handling impasses, an explanation similar in style to certain accounts of children's problems with subtraction (Brown and Burton, 1978).

It is possible, of course, that misconceptions are not always harmful to learning. Indeed, Hook *et al.* (1990) ask whether gaining (and losing) misconceptions actually plays a vital part in the learning process. Making use of Taylor's discourse model, they were able to group the subjects of a small empirical study into three classes: those who produced an account for Prolog's behaviour without straying from this mechanistic level; those who solved an impasse in their understanding of the mechanistic level via a temporary excursion to some other domain in which to reason; and those who abandoned an attempt to provide a mechanistic explanation in favour of some other discourse levels. Learners in the first group often seemed unaware when they had a misconception. In the other groups, there was some evidence that students used their 'misconceptions' to help them, but it is not clear how.

3.2 Reducing misconceptions by exhibiting behaviours

The fact that Prolog's execution-time behaviour is not 'on show' is clearly likely to create misconceptions about what it does. There have therefore been several attempts to construct tools to make Prolog's computational mechanisms more visible to novices. One approach has been to make data flow more visible (Dichev and du Boulay, 1988), but in general attention has been focused on the problem of exhibiting control flow.

Rajan (1986) has outlined a principled design for dynamic tracing environments and created the APT (animated program tracer) system. Among his principles are the following. The code seen in the tracer is to be a direct copy of the code prepared by the novice. (This is fiercer than it sounds: Prolog tracers generally replace variable names with internal representations, making the trace very hard to match up to the original code.) No extraneous symbols are to be inserted, and the code is to be evaluated in the debugging environment in exactly the same order that it would be evaluated in the normal environment. Side-effects are to be made visible, and novices are not to be compelled to learn new command sets and display formats, nor to understand unnecessary multiple viewpoints. In general, Rajan's principles are built round 'What You See Is What Happens', and a respect for the cognitive load experienced by the learner.

In a slightly different approach to the same end, Brayshaw and Eisenstadt (1989) have provided a fine-grained account of Prolog execution that is useful both for teaching and debugging, supported by a graphical tracer/debugger, TPM (the transparent Prolog machine), that allows the execution of a program to be displayed as an 'AORTA' structure (AND-OR tree, augmented). One of the most important aspects of their work is that their goal is a system that is simple enough to be useful to novices, yet powerful enough to be used by professional programmers working with big programs. This means that graphical techniques have to be designed with great care – otherwise they will run out of screen space. At the same time, they wanted to tell the user far more than is available from conventional Prolog tracers, which do little beyond reporting success, failure, and trying again to satisfy a goal. Part of the secret in the AORTA notation is to differentiate between no less than nineteen different types of behaviour by the Prolog interpreter – 'about to attempt new goal', 'system primitive', three types of success, six types of failure, and so on. Each of these has its own compact marking on the tree. Even so, screen space is at a premium. To cope with this, the output has an overall view, with detail suppressed, and a window that can be used to zoom in on any part of the trace that is of interest.

3.3 Supporting plan-level program construction

Evidence on 'programming plans' or schemas is presented elsewhere in this volume, showing that the knowledge of experienced programmers includes knowing sets of instructions to achieve certain familiar small tasks, such as forming a sum over a list. For an excellent account of the programming plans approach to learning programming, see Rist (1989). Although little is yet known about the repertoire of Prolog plans (see Brna *et al.*, 1990, for a review of progress), much can be achieved in this area by armchair consideration of plan structures. Gegg-Harrison (1990) has detailed an approach to teaching students how to build list processing programs from schemas. To have some idea of the complexity of the Prolog world, it is instructive to consider his classification: six simple schemas and eight complex schemas grouped into two different parallel hierarchies according to different viewpoints, with secondary dimensions as well, such as whether the program is a function or a predicate.

The would-be Prolog programmer evidently has a good deal to learn. Learning through a schema-based system may enable much better learning, or it may overwhelm the learner in fine distinctions which are not yet comprehensible at an early stage of learning. This will be an interesting area of development.

4 Programming as the control of real devices: DESMOND and HyperTechnic

All the languages mentioned so far, Logo, Prolog and Solo, enforce new and difficult abstract ideas, such as procedure, parameter and recursion. In response to the resulting problems one school of thought is to move away from the abstract towards the programming of concrete devices. This, indeed, was part of the original Logo platform, the physical turtle that crawled slowly around the floor and could be made to draw lines as it crawled; but Logo still has a high content of syntactic abstractions. The two projects we shall mention have very clearly tried to avoid syntactic load, and have equally clearly tried to supply a very concrete picture so that the user's mental model would be very likely to be highly accurate. Both projects use genuine, physical devices. Unfortunately we have no space to describe explorations in the direction of using highly realistic simulated worlds, such as the 'alternate reality kit' (Smith, 1986) or the 'interactive book' where physics experiments can be programmed by an author and reprogrammed by a user (separately proposed in the 'Boxer' language by diSessa and Abelson, 1986, and in 'Thinglab' by Borning, 1985).

The first project, Desmond, is only one step removed from a good electronics kit – 'Desmond' stands for 'digital electronic system made of nifty devices', and the project was developed at the Open University in association with the 'Microelectronics in Schools' initiative. The kit comprised a small microprocessor, a small touch-sensitive keypad, several sensors (for light, temperature, tilt, etc.) and several output devices (coloured LEDs, a buzzer, a small stepping motor, and a small LCD screen). The microprocessor was programmed in a very conventional single-address assembly code to read sensors and respond to them. Although the maximum program length was quite short, at about 100 steps, a good deal could be achieved. A simple example of a Desmond program fragment is:

```
LDA   91          load accumulator from address 91
ADI   01          add 1 to accumulator
STA   91          store accumulator in address 91
```

This small fragment adds 1 to the value held in location 91.

It can be seen that Desmond code is very much lower level than the other languages considered. There are no special abstractions, such as 'formal parameter', to be mastered; no unexpected syntactic complexities, such as Logo's use of colons and square brackets; no semantic complexities, such as unification; no hidden states of the machine – the entire machine state can be inspected via the LCD screen, which shows the contents of any address in binary, denary and instruction code format. Every instruction has a clear and simple meaning which is unambiguously described in concrete terms of changes to the machine state. How could it fail?

Jones (1989) studied a group of adults learning to use Desmond. They reported many different types of difficulty, some with the teaching materials and some with the device (e.g. expectations that memory-mapped devices, such as the lights, would have the least-significant bit at the left-hand end, whereas it was actually at the right).

Outstanding among their problems were confusions between contents and location – especially in the context of indirect addressing – and difficulties in comprehending the behaviour of a computer address, as something containing a value that is readily overwritten but less-easily put aside. This problem, of course, has afflicted many generations of computer novices. Evidently semantic complexities have *not* been banished, after all!

Jones's studies also pointed to another type of problem: users could see no plans. The role of 'programming plans' in novice learning has been mentioned above; here, the importance of comprehending programs at the functional level of goals and plans was again demonstrated, but this time at the lowest possible language level. Her analysis picks out a number of plans used *by implication* in the teaching materials. The fragment above adds 1 to an address. Other plans that she identified as being used by implication, but not explicitly described, included plans to operate one of the on-board devices, using one device (e.g. switches) to control another (e.g. lamps), adding two numbers, delay loop, test for non-zero numbers (this gave the learners great trouble), bit masking, etc. Readers familiar with assembly code working will recognize many of these from their own experience. But these higher-level plans were not taught as such, and of course the Desmond device contained no tools to identify plans. Learners were left to discover them for themselves, which in some cases was far from easy. Failing to do so meant that program construction became very arduous.

Finally, the Desmond work also points up the importance of visibility and 'viscosity' in programming. The LCD screen gave information about only *one* program location at a time: minimal visibility, making it hard to gain an overview of the program. Similarly, the editor allowed one program location to be changed at a time, and the means for doing so somewhat long-winded: finding the location to be changed and making the change could require many keystrokes. Thus program modification was labour intensive. In the terms used by Green (Chapter 2.2), the poor visibility and high viscosity did not make opportunistic programming possible. Desmond was designed for the 'errorless transcription' view of programming.

As technology improves, so do the opportunities for solving the problems of viscosity and visibility. In an interesting new approach, the resources of HyperCard on the Apple Macintosh are being exploited. Whalley (1990) has adapted the Lego building kits for children, which can be used to make up concrete microworlds such as train lines with level crossings. In his HyperTechnic system, described as 'a graphic object-orientated control environment', a sensor can detect the approach of a train towards the level crossing. The children can tell the sensor what to do when it hears a train coming, such as send a message 'Train coming' to the level crossing. The children can also tell the level crossing what to do when it hears the message 'Train coming': instructions available include go up, go down, wait (which leads to a prompt asking how many seconds), and send a message (see Figure 4).

Whalley reports that 'Trials have shown that children as young as seven can learn to use the train system, and are then able to teach others. An interesting, and as yet unresolved issue, is whether this success is due to the intuitively accessible graphic interface, or to the easy comprehension of the underlying actor language'.

Work of this sort is well ahead of what can be put into the average school. (The trains system requires an Apple Macintosh *and* an Apple II; and even so the microworld is still very limited.) But as a foretaste of the future possibilities, it is startling to see how responsive a system can be built. And yet, one has to apply the

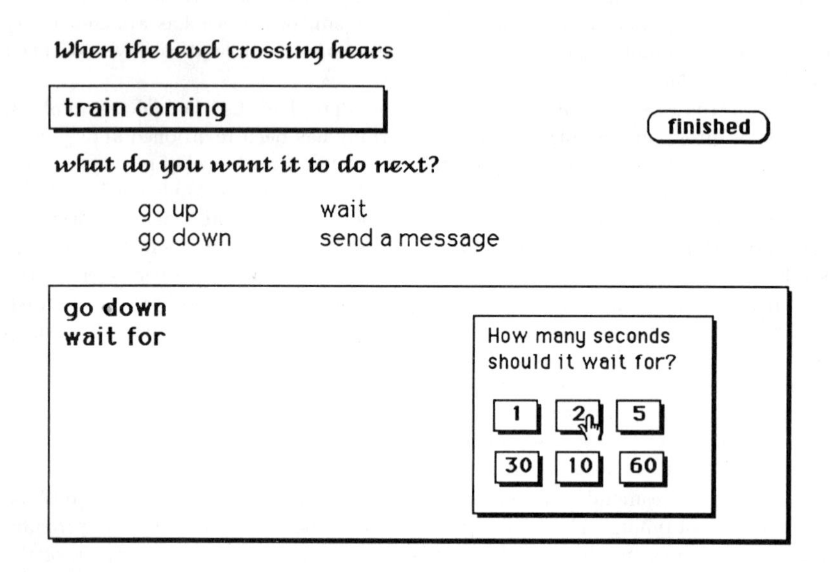

Figure 4: The HyperTechnic programming interface (redrawn from Whalley, 1990).

same criteria. Are global structures visible? No. As HyperTechnic programs grow, they will require some better system to make visible the structures of the program. Moreover, although programming any individual device has been made unbelievably easy, reprogramming the whole structure will remain difficult. But perhaps that is what learning to program is partly about: learning to manage increasing amounts of complexity in a partly abstract system, anticipating as many future contingencies as possible.

5 Matching 'natural plans': BridgeTalk

In this section we present the latest reported version of 'Bridge' and its successor 'BridgeTalk', developed by Bonar and associates (Bonar *et al.*, 1987; Bonar and Liffick, 1990). First we describe the rationale behind their novel and possibly very fruitful approach.

Bonar's thinking starts from programming plans, which have been introduced elsewhere in this volume. Programming plans are exactly what novice programmers lack, according to some authors, and therefore what they need to learn. It is not that novice programmers – even children – are planless; they have perfectly good informal plans for counting sheep, adding up totals, etc. But they do not know how to translate those into programming languages. Conventional programming languages hide programming plans by dispersing them throughout the text, and so novices, instead of learning how to move from informal plans to formalized plans, instead spend time on purely surface features. 'In our video protocols of novice programmers, we see novices working linearly through a program, choosing each statement based on *syntactic features* of the problem text or program code' (Bonar and Liffick, 1990, p. 330, our italics).

On the other hand, Bonar and Liffick argue, the target of teaching programming should be the ability to understand and use a conventional programming language. So we should not seek to evade this responsibility by teaching a simplified programming language which merely postpones the problem of getting to grips with real programming. Instead, the solution is to use an *intermediate representation*, one which minimizes initial difficulty but which leads into real programming. What should that intermediate representation be?

BridgeTalk boldly sets out to encourage *plan-level programming in Pascal*. It is intended to encourage novices to recognize how their informal plans fit into a programming environment, to support them in learning a vocabulary of programming plans, to teach them how to implement those plans in a standard programming language, and to support plan-like composition of programs.

The version of BridgeTalk that we shall describe is no less than the sixth generation. (Readers interested in seeing how the ideas developed should consult Bonar and Liffick, 1990.) Each particular plan has its own icon, which fit together like a jigsaw. Slots in the icons can hold smaller icons for values and constants. Figure 5 shows a program for finding the average of a data set using plan icons; the flow of control is handled by an icon for a loop 'that repeatedly gets a new value until that new value equals a specified sentinel value', a new-value controlled loop plan. 'The key idea with [this plan] is to hide all the syntactic and control flow complexity that a student would need to confront [in order to] implement such a loop in a standard language' (Bonar and Liffick, 1990, p. 338).

One of the most attractive features of BridgeTalk is that it uses graphical representations and mouse dragging to provide a radically new information display, in marked contrast to many ventures into graphics programming (see Chapter 1.2). This representation not only makes both data flow and control flow information available in a well-balanced way, it *also* appears to present plan-level information in a usable manner. Moreover, the environment allows program development to take place in a natural, unforced way. Whereas many tutorial environments use a syntax-based editor to impose a rigid, top-down development on the novice, in order to teach him or her how to 'think properly' (see, for instance, the 'Struedi' Lisp editor for novices described in Chapter 1.2), BridgeTalk eschews such authoritarian pedagogy and allows free construction:

> We have become increasingly suspicious of the glib discussion of 'top-down design' found in most programming textbooks. In particular, the design of BridgeTalk shows that programming design involves many different mappings, including informal to formal, declarative to procedural, goals to plans and processes, natural language to Pascal, linear structure to tree structure, and weakly constrained to strongly constrained. We believe that programming texts do their students a disservice by presenting a design model that at best ignores the differences between novices and experts and at worst is completely unrelated to actual programming practice. (Bonar and Liffick, 1990, p. 363).

We agree.

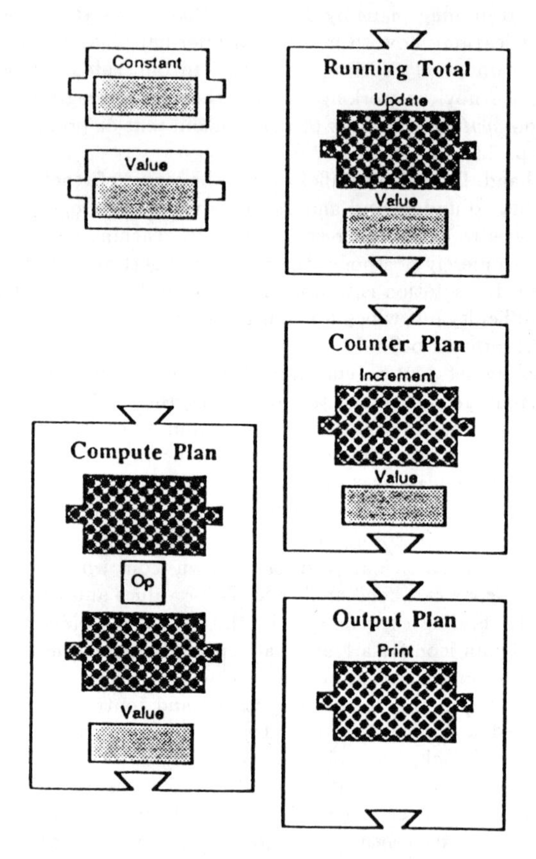

Figure 5: BridgeTalk icons that express a sequence of values, compute new values, and operate on values.

6 Conclusions

This chapter on how to make computing more accessible for educational purposes started with a view of programming as the 'great white hope' of education. Klahr and Carver (1988, p. 363) speak of 'the possibility that it [computer programming] may be the long-sought mental exercise that enables its practitioners to increase their general thinking abilities'. Perhaps there really is such a mental exercise, and just possibly programming is it. But neither anecdotal nor laboratory evidence gives strong support to this hope.

A lot of the learner's time is spent learning the repertoire of programming tricks and techniques, and so our next topic was the comparative lack of success of Prolog, a language that seemingly promised to free the learner of much of that baggage. Experience now suggests that while removing the syntactic apparatus of control flow and data declaration is perhaps a benefit, it also takes away the information about how the program works and thereby mystifies the learner. Contemporary research in Prolog is therefore aiming to create environments that will supply the information about execution which comes freely available with traditional languages.

Also, we glanced briefly at two very different projects which both aim to provide a more concrete learning world for programming, thereby overcoming the difficulties of abstraction and high-level conceptualization in other programming environments. Research on the first of these two, the Desmond system, has shown that despite the concreteness, learners still have problems, and that many of their problems are caused by inability to discover the high-level plans and how to put them together.

The last topic, therefore, was a plan-based system, which consciously attempted to help learners understand the target language, Pascal, in terms of 'everyday plans', via a bridging representation, BridgeTalk. It is far too early to evaluate this system, but the goal is impressive, and so is the care in designing the interface for maximum visibility of important components and for supporting opportunistic planning.

So, where does all this leave us? We present our conclusions in terms of slogans.

Forget about the transfer of competence dispute ...

Disputes about educational objectives are unnecessary and (in the current state of language design) premature. We contrasted 'learning to program' and 'programming to learn'; within the 'learning to program' choice, a programming language is above all a new language to be learnt. Eventually it can serve in the teaching of conventional curricula. Within the alternative choice, 'programming to learn', a language like Logo acts as a medium for the practising of specific skills, such as geometric planning; and it can also be a source of self-knowledge, a mirror for watching oneself solve problems. A language like Prolog can also successfully be used as a 'knowledge bank' in teaching history or ecology – topics where many detailed facts need to be available (e.g. Rasmussen, 1988).

... and go for programming in context.

Alternatively, programming can be used as a step in understanding the world that is coming into being around us, the world of computer applications and computer methods. Soon we shall be living in a world where knowing how to choose, learn and use a computer application will be of the first importance – a world where many of the tasks will be either quite new, or else greatly affected by the existence of computer systems. By programming even the most elementary versions of some of these applications, whether a word processor, a database management system, or a circuit design package, an inside view will have been obtained.

Make it easy to comprehend ...

Not many educational objectives can be accomplished if would-be learners are frustrated by unnecessary difficulties with the programming language and environment. It seems to the authors that far too much attention has been given to the debate on transfer of competence, and far too little to attacking the problems of ease of learning and ease of use.

Back in 1981, du Boulay *et al.* described the difficulties in presenting computing concepts to novices. They introduce the concept of the 'notional machine', '... an idealized, conceptual computer whose properties are implied by the constructs in the programming language employed'. This notional machine should be kept *functionally* simple by giving it a small set of what they call 'transactions' (after Mayer, 1979); it should be kept *logically* simple, so that problems of interest to the novice can be tackled by short, simple programs; and it should be *syntactically* simple, i.e. the rules for writing instructions should be uniform and have well-chosen names. But simplicity is not enough. The notional machine must also be *visible in its own terms*. This is the crux of the problem in many cases – novices cannot see what the notional machine is doing.

Today it is clear that comprehensibility requires more than simplicity and visibility. Jones (1990), in her study of Desmond, Logo, Solo, and a microprocessor language not described here, showed that even in systems meeting those criteria, problems still arose. She demonstrated the existence of unexpected semantic complexities and the need to help learners understand at the functional level, the level of plans and overviews.

... and easy to use ...

It must be obvious to anybody that systems for novices should be easy to use. Yet novices are still struggling with inadequate systems! Too much typing, with no spelling correction; too much syntax, with poor error messages and no error correction or 'Do What I Mean'; poor visibility of different parts of the program, and poor display of execution behaviour – these are still typical. Perhaps one reason has been the habit of treating the programming language and its operating environment in very separate ways. Basic, for all its faults, supplied a seamless environment for learners; one wishes every innovator had done the same. DiSessa and Abelson (1986) observe that 'User interfaces are often considered to be separable from programming language semantics and almost an afterthought in language design. Worse, most present languages assume only character-stream input and output. A useful medium must be much more flexibly interactive'.

This persistent separation of language and environment reflects slow acceptance of the ideas of opportunistic planning and exploratory programming. Exploratory programming requires a suitable support environment, of course. Chapter 2.2 ('Programming languages as information structures') shows how recently this idea has displaced previous views, notably the view of programming as 'errorless transcription'.

... and avoid the dogmas of economy.

Finally, it must be remembered at all times that the users are *learners*. Their needs, their knowledge and their abilities are different from those of experienced programmers, who may well have been self-selected for some rather special attributes. Bonar and Liffick (1990) say in some detail why they think current programming languages are unsuitable for novices. They point out that their 'emphasis in economy of expression ... is misplaced in designing languages for novice programmers. By looking for ever more abstracted ways to express similar procedural behavior, modern languages have excised most clues to goal and purpose that are essential to novice understanding of a program'. They suggest that students would prefer a language in which each plan was expressed by a different construct. Conversely, modern languages emphasize tools for abstraction, which novices are not yet ready to use.

Once again, we agree with Bonar and Liffick. We look forward to seeing the next generation of languages and environments for easy access by learners.

References

Anzai, Y. and Uesato, Y. (1982). Learning recursive procedures by middleschool children. *Proceedings of the Fourth Annual Conference of the Cognitive Science Society.* Ann Arbor.

Borning, A. (1985). A prototype electronic encyclopaedia. *ÈACM Transactions on Office Information Systems,* **3**, 63-88.

Bonar, J. and Liffick, B. W. (1990). A visual programming language for novices. *In* S.-K. Chang (Ed.), *Principles of Visual Programming Systems.* Englewood Cliffs: Prentice-Hall.

Bonar, J., Cunningham, R., Beatty, P. and Riggs, P. (1987). Bridge: intelligent tutoring with intermediate representations. Technical Report, Learning Research and Development Center, University of Pittsburgh.

Bourbion, M. (1986). *Le choix Logo.* Paris: Armand Colin Editeur.

Brayshaw, M. and Eisenstadt, M. (1989). A practical tracer for Prolog. Technical Report no 42, Human Cognition Research Laboratory, Open University, Milton Keynes. *International Journal of Man-Machine Studies*, in press.

Brown, J.S. and Burton, R. (1978). Diagnostic models for procedural bugs in basic mathematical skills. *Cognitive Science,* **2**, 155-192.

Bundy, A. (1984). What stories should we tell Prolog students? Working Paper 156, Department of Artificial Intelligence, Edinburgh.

Coombs, M. J. and Alty, J. (1984). Expert systems: an alternative paradigm. *In* M. J. Coombs (Ed.), *Developments in Expert Systems.* London: Academic Press.

Crahay, M. (1987). Logo, un environnement propice à la pensée procédurale. *Revue Française de Pédagogie,* **80**, 37-56.

Dichev, C. and du Boulay, B. (1988). A data tracing system for Prolog novices. *In* T. O'Shea and V. Sgurev (Eds), *Artificial Intelligence III: Methodology, Systems, Applications.* Amsterdam: North-Holland.

diSessa, A. A. and Abelson, H. (1986). Boxer: a reconstructible computational medium. *Communications of the ACM,* **29**, 859-868.

du Boulay, B., O'Shea, T. and Monk, J. (1981). The glass box inside the black box: presenting computing concepts to novices. *International Journal of Man-Machine Studies,* **14**, 237-249.

Eisenstadt, M. (1979). A friendly software environment for psychology students. *AISB Quarterly,* **34.**

Eisenstadt, M. and Lewis, M. (1985). Errors in an interactive programming environment: causes and cures. Technical Report No. 4, Human Cognition Research Laboratory, The Open University, Milron Keynes.

Fung, P., du Boulay, B. and Elsom-Cook, M. (1987). An initial taxonomy of novices' misconceptions of the Prolog interpreter. CITE Report 27, Centre for Information Technology in Education, Institute for Educational Technology, The Open University.

Gegg-Harrison, T. S. (1990). Learning Prolog in a schema-based environment. *Instructional Science,* in press.

Hofstadter, D. (1979). *Gödel, Escher, Bach: an Eternal Golden Braid.* New York: Basic Books.

Hook, K., Taylor, J. and du Boulay, B. (1990). Redo 'try once and pass': the influence of complexity and graphical notation on novices' understanding of Prolog. *Instructional Science,* in press.

Howe, J.A.M. and O'Shea, T. (1978). Computational metaphors for children. *In* F. Klix (Ed.), *Human and Artificial Intelligence.* Berlin: Deutscher Verlag.

Hoyles, C. and Noss R. (1987). Synthesising mathematical conceptions and their formalisation through the construction of a Logo based school mathematics curriculum. *International Journal of Mathematics Education in Science and Technology,* **18.**

Jones, A. (1989). Empirical studies of novices learning programming. Ph.D. thesis, Institute of Educational Technology, The Open University, Milton Keynes.

Kahney, H. (1983). Problem solving by novice programmers. *In* T.R.G. Green, S.J. Payne and G.C. van der Veer (Eds), *The Psychology of Computer Use.* London: Academic Press. Reprinted in E. Soloway, and J. C. Spohrer, (Eds), *Studying the Novice Programmer.* Hillsdale, NJ: Erlbaum.

Kessler, C.M. and Anderson, J.R. (1986). Learning flow of control: recursive and iterative procedures. *Human Computer Interaction,* **2,** 135-166.

Klahr, D. and Carver, S. M. (1988). Cognitive objectives in a Logo debugging curriculum: instruction, learning, and transfer. *Cognitive Psychology,* **20,** 362-404.

Kurland, D.M., Pea, R.D., Clement, C. and Mawby, R. (1989). A study of the development of programming ability and thinking skills in high school students. *In* E. Soloway and J.C. Spohrer (Eds), *Studying the Novice Programmer.* Hillsdale, NJ: Erlbaum.

Lawler, R.W. (1985). *Computer Experience and Cognitive Development: A Child's Learning in a Computer Culture.* Chichester: Ellis Horwood.

Lehrer, R., Guckenberg, T. and Sancilio, L. (1988). Influences of Logo on children's intellectual development. *In* R.E. Mayer (Ed.), *Teaching and Learning Computer Programming.* Hillsdale, NJ: Erlbaum.

Littlefield, J., Delclos, V.R., Lever, S., Clayton, K.N., Brandsford, J.D. and Franks J.J. (1988). Learning Logo: method of teaching, transfer of general skills, and attitudes toward school and computers. *In* R.E. Mayer (Ed.), *Teaching and Learning Computer Programming.* Hillsdale, NJ: Erlbaum.

Mayer, R.E. (1979). A psychology of learning Basic. *Communications of the ACM*, **22**, 589-593.

Mendelsohn, P. (1985). Learning recursive procedures through Logo programming. *Proceedings of the Second Logo and Mathematics Education Conference.* University of London.

Mendelsohn, P. (1988). Les activités de programmation chez l'enfant: le point de vue de la psychologie cognitive. *Technique et Science Informatiques*, **7**, 47-58.

Nichol, J., Briggs, J. and Dean, J. (Eds) (1988). *Prolog, Children and Students.* London: Kogan Page.

Ormerod, T. C., Manktelow, K. I., Robson, E. H. and Steward, A. P. (1986). Content and representation effects in reasoning tasks in Prolog form. *Behaviour and Information Technology*, **5**, 157-168.

O'Shea, T. and Self, J. (1983). *Learning and Teaching with Computers: Artificial Intelligence in Education.* Brighton: The Harvester Press.

Pain, H. and Bundy, A. (1987). What stories should we tell novice Prolog programmers. *In* R. Hawley (Ed.), *Artificial Intelligence Programming Environments.* Chichester: Ellis Horwood.

Papert, S. (1980). *Mindstorms: Children, Computers and Powerful Ideas.* New York: Basic Books.

Pea, R.D. and Kurland, D.M. (1984). On the cognitive effects of learning computer programming. *New Ideas in Psychology*, **2**, 137-168.

Pirolli, P. (1986). A cognitive model and computer tutor for programming recursion. *Human-Computer Interaction*, **2**, 319-355.

Putnam, R.T., Sleeman, D., Baxter, J.A. and Kuspa, L.K. (1986). A summary of misconceptions of high school Basic programmers. *Journal of Educational Computing Research*, **2**, 57-73.

Rajan, T. (1986). APT: A principled design for an animated view of program execution for novice programmers. Technical Report 19, Human Cognition Research Laboratory, The Open University.

Rasmussen, J. (1988). Using Prolog in the teaching of ecology. *In* J. Nichol, J. Briggs and J. Dean (Eds), *Prolog, Children and Students.* London: Kogan Page.

Rist, R. (1989). Schema creation in programming. *Cognitive Science*, **13**, 389-414.

Roberts, E.S. (1986). *Thinking Recursively.* New York: Wiley.

Rogalski, J. (1987). Acquisition et didactique des structures conditionnelles en programmation informatique. *Psychologie Française*, **32**, 275-280.

Rogalski, J. and Vergnaud, G. (1987). Didactique de l'informatique et acquisitions cognitives en programmation. *Psychologie Française*, **32**, 267-274.

Ross, P. and Howe, J. (1981). Teaching mathematics through programming: ten years on. *In* R. Lewis and D. Tagg (Eds), *Computers in Education.* Amsterdam: North-Holland.

Rouchier, A. (1987). L'écriture et l'interprétation de procédures récursives en Logo. *Psychologie Française,* **32,** 281-285.

Smith, R. (1986). The alternate reality kit: an animated environment for creating interactive simulations. *Proceedings of the 1986 IEEE Computer Society Workshop on Visual Languages, Dallas.* IEEE.

Solomon, C. (1986). *Computer Environments for Children: A Reflection on Theories of Learning and Education.* MIT Press.

Soloway, E., Bonar, J. and Ehrlich, K. (1983). Cognitive strategies and looping constructs: an empirical study. *Communications of the ACM,* **26,** 853-860.

Suppes, P. (1979). Current trends in computer-assisted instruction. *In* M.C. Yovits (Ed.), *Advances in Computers,* vol. 18, New York: Academic Press.

Taylor, J. (1987). Programming in Prolog: an in-depth study of problems for beginners learning to program in Prolog. Unpublished Ph.D. thesis, School of Cognitive Studies, University of Sussex.

van Someren, M. W. (1985). Beginners' problems in learning Prolog. Memorandum 54, Department of Experimental Psychology, University of Amsterdam.

van Someren, M. W. (1988). What's wrong? Understanding beginners' problems with Prolog. VF Memo 89, Department of Social Science Informatics, University of Amsterdam.

van Someren, M.W. (1990). Understanding students' errors with Prolog unification. *Instructional Science,* in press.

Vitale, B. (1987). Epistemology and pedagogy of children's approach to informatics. *Proceedings of the International Conference on Education.* Bilbao.

Whalley, P. (1990). HyperTechnic – a graphic object-orientated control language. *Proceedings of the 7th Conference on Technology and Education.* Brussels, 1990.

White, R.H. (1988). Effects of Pascal knowledge on novice Prolog programmers. Research Paper 399, Department of Artificial Intelligence, Edinburgh.

Part 3

Expert Programming Skills and Job Aids

After having examined programming language features, their learning by beginners, their implications on the structuring of programming knowledge, and their use in education (Part 2), this part is devoted to programming at an expert level. The perspective is wider, less dependent on programming languages than that of the previous part, and relevant to other design activities.

As shown in Part 1 of this volume (Chapter 1.3), programming combines several features which together form expert skills. Two basic skills have received much attention at an expert level: program understanding and designing. These are basic in that they are essential components of other skills. For example, program debugging is closely linked to program understanding: experts elaborate a hierarchical representation of the program control structure and function before beginning to detect or correct errors (Vessey, 1985, 1989). Programmers form a mental reconstruction of what a 'correct' program is, to compare it to the actual program (Michard, 1975); hence the close link between program debugging and program designing.

The three chapters in Part 3 exemplify the need to decompose a complex programming activity in order to gain a better understanding of its components. This carries the concomitant risk of transforming these components in the simplified situations under study. However, Chapter 3.3 and Part 4 address the question of 'programming in the large' in more complex and everyday work situations. The chapters in Part 3 shed light on complementary features of programming, rather than taking exclusively theoretical positions.

The first two chapters present two kinds of approaches to program understanding, and reflect the fundamental duality of cognition calling for use of declarative and procedural knowledge at the same time (see Chapter 1.4). Program-understanding is studied, for example using verbal reports on expectations during program reading, or in the context of debugging or recall tasks, etc. Some data analysis techniques are more directed to identifying expert declarative knowledge bases, especially program plans and control structures used in program understanding (Détienne, Chapter 3.1). Other techniques are needed to investigate understanding strategies more directly, and to examine the respective role of expert knowledge bases and expert strategies in the program understanding expertise (Gilmore, Chapter 3.2). Implications in terms of programming teaching touch on the importance of teaching both kinds of knowledge.

Aside from these investigations into program understanding, studies have been devoted to expert design strategies. The last chapter of Part 3 reviews these studies, presents dimensions used to classify strategies and some recommendations for developing job aids (Visser and Hoc, Chapter 3.3). In this chapter, programming is considered as a particular case of a design activity and the framework and recommendations go beyond the sole computer programming.

References

Michard, A. (1975). Analyse du travail de diagnostic d'erreurs logiques dans un programme FORTRAN. Le Chesnay (F.), INRIA, Research Report no. C07602R48.

Vessey, I. (1985). Expertise in debugging computer programs: a process analysis. *International Journal of Man-Machine Studies*, **23**, 459-494.

Vessey, I. (1989). Toward a theory of computer program bugs: an empirical test. *International Journal of Man-Machine Studies*, **30**, 23-46.

Chapter 3.1

Expert Programming Knowledge: A Schema-based Approach

Françoise Détienne

Projet de Psychologie Ergonomique pour l'Informatique, INRIA, Rocquencourt, BP 105, 78153, Le Chesnay Cedex, France

Abstract

The topic of this chapter is the role of expert programming knowledge in comprehension. In the 'schema-based approach', the role of semantic structures is emphasized whereas, in the 'control-flow approach', the role of syntactic structures is emphasized. Data that support schema-based models of understanding are presented. Data that are more consistent with the 'control-flow approach' suggest limitations of the former kind of models.

1 Introduction

The structures of knowledge and its organization is an important characteristic of a model of the expert programmer. Having such a model enables the design of systems which support programming activities. Such a model is also necessary to understand how expertise is acquired in a domain. The topic developed in this chapter is the role of expert knowledge in program comprehension. This discussion is limited to comprehension, because first, as shown in Chapter 1.3, the same kind of knowledge

Psychology of Programming
ISBN 0-12-350772-3

is used by the processes of program composition and program comprehension, and second, the understanding activity is particularly interesting to analyse because it is involved in several maintenance tasks. Constructing a mental representation of a program is necessary for effectively debugging or modifying a program.

Many studies on programming knowledge have been conducted in the theoretical framework of schema theory (see Chapter 1.4 for details of this theory). This approach is developed here and is evaluated on the basis of empirical data.

In Section 2, studies on program comprehension are briefly reviewed from a historical perspective. In the 'schema-based approach' the role of semantic structures is emphasized whereas, in the 'control-flow approach' the role of syntactic structures is emphasized. In Section 3, a theory of the knowledge structures which programmers are supposed to possess is developed. Then, empirical support for these hypotheses on knowledge structures is presented. In Section 4, comprehension mechanisms are analysed through empirical results.

2 A Historical perspective on approaches to program comprehension

Different approaches to comprehension can be distinguished according to the knowledge and the cognitive mechanisms that are assumed to be used in experts' program comprehension activities. There is disagreement as to the kind of knowledge that is important when understanding a program.

In the schema-based approach experts are assumed to use mainly knowledge structures that represent semantic information in the programming domain. These structures group together information on those parts of programs that perform the same function. In the control-flow approach experts are assumed to use mainly knowledge structures that represent elements of the control structure. These structures group together those parts of programs occurring within the same syntactic structure. The former emphasizes the role of semantic knowledge, i.e. knowledge about what the program does whereas the latter emphasizes the role of syntactic knowledge, i.e. knowledge of how the program works.

From a theoretical viewpoint, proponents of the schema-based approach have used concepts developed in the schema theory. Knowledge structures in programming are formalized in terms of 'programming schemas' or 'programming plans'. Schema theory is not always referred to in the control-flow approach. However, the schema concept could also be used to model the syntactic knowledge structures that experts are assumed to possess.

The schema theory is a theory of knowledge organization in memory and of the processes involved in using knowledge. A schema is a data structure that represents generic concepts stored in memory. This theory has been developed in artificial intelligence (Abelson, 1981; Minsky, 1975; Schank and Abelson, 1977) and in psychological studies on text understanding (Bower *et al.*, 1979; Rumelhart, 1981).

The schema-based approach in programming studies began with Rich's work (1981) and has been mainly developed by the group headed by Soloway. It can account for problem solving and understanding activities in various programming tasks such as program design, debugging and enhancement. The studies started in the early 1980s and some of the most important papers on this topic are by Soloway *et al.* (1982a,b) and Soloway and Ehrlich (1984). For several years, other researchers from different institutions (Brooks, 1983; Détienne, 1989; Rist, 1986) have conducted studies on programming using the same theoretical framework. In the control-flow approach, studies have also been performed by both computer scientists and psychologists (Atwood and Ramsey, 1978; Curtis *et al.*, 1984; Linger *et al.*, 1979; Pennington, 1987; Shneiderman, 1980).

There is not only disagreement on the kind of knowledge that is important for comprehending a program, but also on the way knowledge is used in program comprehension. On one side, proponents of the schema-based approach assume that understanding a program consists of the evocation of a programming schema (or several schemas) stored in memory, instantiating that schema with values extracted from the text and inferring other values on the basis of the evoked schema. The mechanisms of schema activation can be either data driven or conceptually driven. In the first case the activation spreads from substructures to superstructures, and, in the second case, the activation spreads from the superstructures to the substructures. Therefore, semantic structures can be evoked directly by information extracted from the code or by other activated schemas.

On the other side, proponents of the control-flow approach assume that understanding consists in identifying control structures and then combining these structures into larger structures until all structures correspond to some function. A good example of this approach is the syntactic/semantic model of program comprehension, developed by Shneiderman and Mayer (described by Shneiderman, 1980). This assumes that comprehension involves two separate processes, with syntactic processing occurring first followed by semantic processing.

In the rest of this chapter, the schema-based approach is developed. Data that support schema-based models of understanding are presented. However, data that are more consistent with the control-flow approach are also presented, suggesting some limits to schema-based models.

3 Knowledge organization in memory

3.1 Theoretical framework

In the schema-based approach to comprehension, hypotheses are made on the typology of schemas possessed by experts, the relationship between schemas, and the structure of representations constructed from programs. These points are developed in this section. Remarks are also made on the knowledge organization assumed in the control-flow approach.

3.1.1 Typology of schemas possessed by experts

Soloway *et al.* (1982a) have encoded experts' schema knowledge as frames. A schema is represented as a knowledge packet with a rich internal structure. A schema (or 'plan' in their terminology)[1] is composed of variables ('slot types') that can be instantiated with values ('slot fillers').

Program understanding involves the evocation of schemas of different domains and their articulation. Brooks (1983) assumes that program understanding involves a mapping between, at least, two domains, the programming domain and the problem

[1] I will use the terms 'plan' and 'schema' interchangeably throughout this chapter.

domain. Experts would possess schemas representing information on problems which are dependent on the problem domain (or task domain). Détienne (1986b) describes the general structure of those schemas as including slots on the data structure and on the possible functions in a particular problem domain. For the particular domain of stock management, the following values could be assigned to the slots:

Task domain: stock management

Data structure: record (name of file, descriptor of file.)

Functions: allocation (creation or insertion), destruction, search

Among the schemas relative to the programming domain, Soloway *et al.* distinguish variable plans and control-flow plans. Variable plans generate a result that is stored in a variable. For example, the Counter_Variable plan can be formalized as:

Description: counts occurrences of an action

Initialization: Counter := 0

How used? (or update): Counter := Counter + 1

Type: integer

Context: iteration

Control-flow plans do not generate results, rather they regulate the use and production of data by other schemas. For example, the Running_Total_Loop plan, which computes the sum of a set of numbers, initializes a total variable to zero, gets a number, adds that number to its total and loops until a stopping value is entered. It is described by Soloway as the following frame:

Description: build up a running total in a loop, optionally counting the numbers of iterations

Variables: Counter, Running_Total, and New_Value

Set up: initialize variables

Action in body: Read, Count, Total

In this schema, the Running_Total variable refers to a variable that builds up a value one step at a time, like, for example, a variable in which would be accumulated the sum of the numbers. The New_Value variable holds the values produced by the generator. This can be, for example, a Read_Variable plan that receives and holds a newly read variable or a Counter_Variable plan that counts occurrences of an action. Experts would also possess more complex schemas representing algorithms like search or sort algorithms or abstract types like tree structures or record structures.

Pennington (1987) describes different kinds of relations that formally compose a program: control-flow relations that reflect the execution sequence of a program, data-flow relations that reflect the series of transformations on data objects in a program, functional relations that concern the goal achieved in a program. To contrast the schema-based approach and the control-flow approach, Pennington remarks that the plan knowledge analysis is closely related to an analysis of program text in terms of data-flow and functional relations. Programming schemas represent information relative both to functions performed in a program (e.g. search for an item) and to data flow relations. In contrast, the control-flow knowledge analysis is more closely related to elements of the control structures such as sequences, iterations and conditional structures. These basic structures of programs are called 'primes structures' by Linger *et al.* (1979).

Soloway *et al.* assume that experts also possess rules of discourse that are programming conventions about how to use and compose programming plans. For example, a rule might express the convention that the name of a variable should reflect its function. The application of these rules produces prototypical values associated with plans.

3.1.2 Relationship between schemas

Soloway *et al.* (1982b) describe two kinds of relationship between plans: a relation of specialization ('a kind of') and a relation of implementation ('use').

The relation of specialization links together plans that are more or less abstract. A specialized plan forms a subcategory of the plan just above it in the hierarchy. For example a New_Value_Variable Plan which holds values produced by a generator can be specialized in two distinct plans: a Read_Variable plan and a Counter_Variable plan. In the same way, the Running_Total_Loop plan can be specialized as three distinct plans: Total_Controlled_Running_Total Loop plan in which the Running_Total is tested, a Counter_Controlled Running_Total Loop plan in which the Counter is tested and a New_Value_Controlled_Running_Total_Loop plan in which the New_Value is tested.

The relation of implementation links together plans that are language independent on one side, with plans that are language dependent on the other side. Implementation plans specify language-dependent techniques for realizing tactical and strategic plans. For example, the For_Loop plan is a technique for implementing the Counter_Controlled Running_Total_Loop plan in Pascal.

The relation of implementation refers also to the way some plans are composed of elementary plans. For example a Running_Total_Loop Plan is composed of different Variable plans which are associated with it by the use link.

3.1.3 Program representation

In the schema-based approach, a program may be represented hierarchically as goals and subgoals with potential programming schemas associated with each goal or subgoal. Figure 1a illustrates the notions of goals and subgoals for a program that computes the average of a set of numbers. Goals are decomposed into subgoals. For example, the goal 'report-average' is decomposed in three subgoals: 'enter-data', 'compute-average', 'output-average'. At each terminal subgoal in the tree there is an associated schema for coding that subgoal in Pascal.

Figure 1b illustrates the implementation of schemas in a program. It shows, for example, that the variable whose name is 'Count' implements a Counter_Variable

(a)

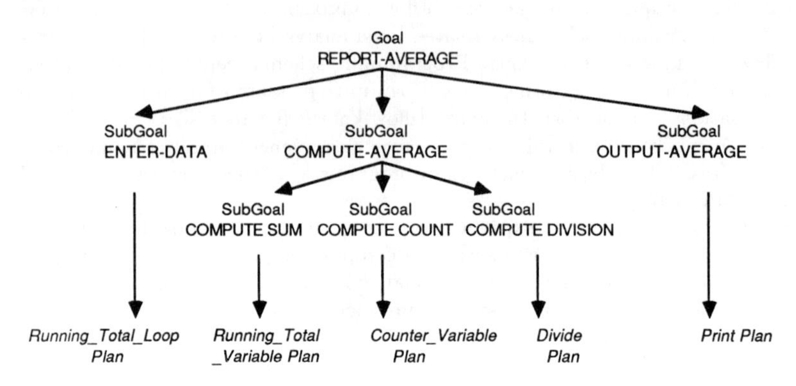

(b)

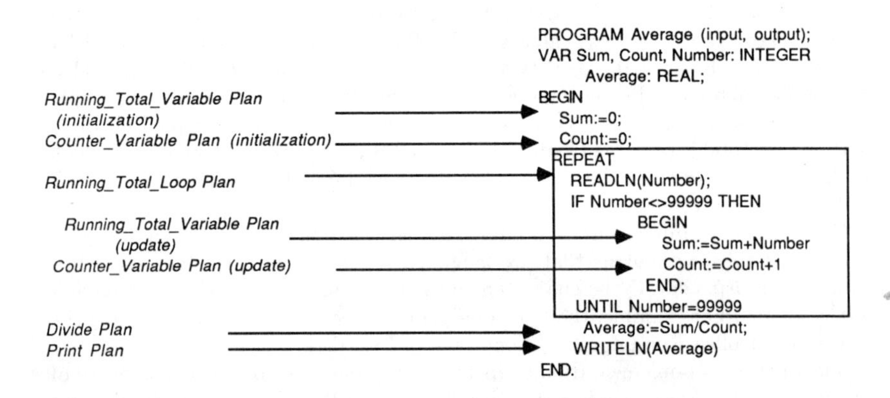

Figure 1: Representations of a program computing an average.
(a) Hierarchical representation of goals and schemas.
(b) Representation of the combination of schemas implemented in the program code.

plan. Its initialization corresponds to the line 'Count := 0' and its update corresponds to the line 'Count := Count + 1'.

In the control-flow approach, a program representation would be structured, at a detailed level, in terms of syntactic structures that could be combined, at a more abstract level, in terms of semantic structures.

3.2 Empirical support

Empirical data from a variety of tasks support the hypothesis that experts possess schemas of programming that represent semantic information. These include categorization tasks and using fill-in-the-blank tasks for plan-like and unplan-like programs.

3.2.1 Categorization of programs

The experts' knowledge organization gives them a capacity of processing that is superior to novices. Experts recall programs better than novices whenever the order of presentation is correct (Shneiderman, 1976) but this superiority disappears when programs are presented in random order so that meaningful structures are not visible.

The categories formed by experts indicate the knowledge structures they use. The categories formed by novices should be different from those formed by experts inasmuch as novices do not possess the same knowledge structures. Using a recall procedure, Adelson (1981) showed that categorization of programs or parts of programs is different according to the expertise of subjects. Novices' categories depend on surface features of the program like syntactic structure, whilst experts' categories cluster around elements performing the same function and elements displaying a procedural similarity.

These results suggest that experts possess knowledge structures grouping together information relative to the same function. This supports the schema-based hypothesis for programming plans. However, Adelson's results also suggest that experts possess knowledge structures that reflect the control structure of the program. This last finding is more consistent with the control-flow approach.

3.2.2 Understanding plan-like and unplan-like programs

Soloway and Ehrlich's results (1984) support the hypothesis that experts possess and use programming plans and rules of discourse in comprehending programs. If this hypothesis is correct, then giving experts programs that have a disrupted plan structure should make them much more difficult to understand than programs which are normal. Novices, who do not yet possess programming plans and conventions, should not be sensitive to whether programs do or do not conform to programming plans.

To evaluate this hypothesis, Soloway and Ehrlich constructed two versions of each of a set of programs: plan-like and unplan-like. In unplan-like versions, either the composition of the plans is not prototypical or the plans are implemented with values that are not prototypical. A further way of constructing unplan-like programs is to violate some rules of programming discourse. Figure 2 presents an example of plan-like and unplan-like versions of a program that computes an average.

PLAN-LIKE VERSION

```
PROGRAM Grey (input,output)
VAR Sum, Count, Num: INTEGER ;
  Average: REAL ;
BEGIN
  Sum:=0;
  Count:=0; * line to fill in
  REPEAT
    READLN(Num);
    IF Num <> 99999 THEN
                                  BEGIN
                                    Sum:=Sum+Num;
                                    Count:=Count+1;
                                  END;
  UNTIL Num=99999;
  Average:=Sum/Count;
  WRITELN(Average)
END.
```

UNPLAN-LIKE VERSION

```
PROGRAM Orange(input,output)
VAR Sum, Count, Num: INTEGER
  Average: REAL ;
BEGIN
  Sum:=99999;
  Count:=-1; * line to fill in
  REPEAT
    READLN(Num);
    Sum:=Sum+Num;
    Count:=Count+1;
  UNTIL Num=99999;
  Average:=Sum/Count;
  WRITELN(Average)
END.
```

DESCRIPTION (extract from Soloway and Ehrlich, 1984)
This program calculates the average of some numbers that are read in; the stopping condition is the reading of the sentinel value, 99999.
The plan-like version accomplishes the task in a typical fashion: variables are initialized to 0, a read-a-value/process-a-value loop is used to accumulate the running total, and the average is calculated after the sentinel has been read.
The unplan-like version was generated from the plan-like version by violating a rule of discourse: *don't do double duty in a non-obvious way.* That is, in the unplan-like version, unlike in the plan-like version, the initialization actions of the COUNTER VARIABLE (Count) and RUNNING TOTAL VARIABLE PLANs (Sum) serve two purposes:
-Sum and Count are given initial values
-the values are chosen to compensate for the fact that the loop is poorly constructed and will result in an off-by-one bug: the final sentinel value (99999) will be incorrectly added into the RUNNING TOTAL VARIABLE, Sum, and the COUNTER VARIABLE, Count, will also be incorrectly updated.

Figure 2: Program AVERAGE.

The subjects performed a fill-in-the-blank task: one line of the code was deleted and the subjects had to fill in the blank line with a line of code that in their opinion best completed the program. They were not told what the program was supposed to do. The results were that the experts performed better than the novices, that all subjects were more likely to get the plan-like version correct more often than the unplan-like versions, but that this difference was much greater for experts than for novices. These results support the hypothesis on experts' plan knowledge structures.

Furthermore, the schema-based model predicts the kind of errors that experts might make in the unplan-like condition. If their understanding is based on programming plans and rules of discourse then they will try to infer the missing line on this basis. Thus, they should tend to give plan-like answers even for unplan-like versions. This was indeed Soloway and Ehrlich's observation.

4 Comprehension mechanisms

4.1 Theoretical framework

According to a schema-based approach of understanding, schemas representing semantic knowledge are evoked while reading a program. They may be evoked either in a bottom-up way or in a top-down way. The direction of activation is bottom-up when the extraction of cues from the code allows the activation of schemas or when an evoked schema causes the activation of schemas it is part of. The direction of activation is top-down when evoked schemas cause the activation of less-abstract schemas; activation spreads from superstructures toward substructures. In Brooks' model (1983), the activation process is assumed to be mostly conceptually driven.

Concerning the evocation of schemas, Brooks attributes an important role to 'beacons', that is, to features or details visible in the program or the documentation as typical indicators of the use of a particular operation or structure. They allow the activation or recognition of particular schemas.

Détienne (1989) stresses that as most schema-based models focus on the construction of goal/plan representation based on activation and instantiation processes, it seems important to analyse the processes that evaluate this representation. Détienne distinguishes two processes: the evaluation of the internal coherence between plans and the evaluation of the external coherence between plans. The former consists in checking whether or not the values instantiated in a plan satisfy the constraints on the instantiation of the plan's slots. The latter consists in checking whether or not there are interactions between plans and between goals and if they create any constraints on the implementation of plans.

In the schema-based approach, the programmer is assumed to construct a representation of the functions performed in the program, making more or less explicit the data-flow relations in the program.

In the control-flow approach, syntactic constructs are assumed to be evoked first. This means that the programmer will construct representations of the control-flow relations in the program before the performed functions. Syntactic constructs are assumed to be combined in more and more abstract constructs until reaching a level of functional representation.

4.2 Empirical support

Data on the inferences a programmer makes in comprehension and recall tasks reflect the kind of knowledge used during comprehension and provide information on the way that knowledge is used. 'Chunks' of program code perceived by a programmer group together information belonging to the same mental constructs. Empirical data supply information on the kinds of chunks constructed by experts. The kind of evoked knowledge and comprehension mechanisms may vary according to the task performed. Some observations support the schema-based approach whereas others support the control-based approach.

4.2.1 Inferences collected in understanding tasks

According to the schema-based approach, schemas representing semantic knowledge are evoked in program comprehension. These schemas permit the drawing of inferences. The inferences collected during comprehension reflect the kind of knowledge that has been evoked in memory. To study this process of schema activation, Détienne (1986a, 1988) designed an experimental setting in which experienced programmers had to verbalize while reading programs (written in Pascal) presented one instruction at a time. At each instruction newly presented, the subjects had to express the information provided by it and elaborate any hypotheses concerning other instructions or the function of the program.

The results supported the hypothesis that experts possess programming plans like those formalized by Soloway. The verbal and behavioural protocols were coded in the form of production rules: 'IF A, THEN B' representing various mechanisms for using knowledge when drawing inferences. Evoked knowledge structures have been formalized as frames 'SCHEMAn' composed of variables 'VARn'. Two hundred and fifty-nine rules have been identified that have been classified as examples of forty-seven general rules. In its general form, a rule describes the types of variables (slot types) composing a schema. In its instantiated form, it describes possible values (slot fillers) that can instantiate a variable. In this section, several examples of these rules are presented in their instantiated form, with the slot types in parentheses.

Some identified rules describe activation processes that are data driven (the direction of activation is bottom up); they are expressed as 'IF VAR1a, THEN SCHEMA1' (VAR1a is a variable composing SCHEMA1) or as 'IF SCHEMA1a, THEN SCHEMA1' (SCHEMA1a is linked to a variable composing SCHEMA1). As instantiation begins as soon as a schema is evoked, some rules also express instantiation processes; expressed as 'IF VAR1a, THEN VAR1b' (VAR1a and VAR1b are two variables that are part of the same schema).

Détienne observed that a Counter_Variable plan, for example, can be evoked by the extraction of cues such as the variable's name or its form of initialization. This is illustrated by the following rules:

IF VAR1a (variable's name): I
IF VAR1b (variable's type): integer
THEN SCHEMA1 (schema of variable): Counter_Variable plan
THEN VAR1c (context): iteration

IF VAR1a (variable's initialization form): I:= 1
THEN SCHEMA1 (schema of variable): Counter_Variable plan
THEN VAR1b (variable's update): I:= I+ 1

These rules make it clear that the experts infer other values associated to a Counter_Variable plan when it is activated. According to the first rule, the reading of the declaration of a variable whose name is 'I' and type is 'integer' evokes a Counter-variable plan. This activation allows the programmer to infer the value associated with another slot of this schema which is the context in which the variable 'I' is used. Thus, the subject will expect to see some kind of iteration in the program code. According to the second rule, the activation of a Counter-variable plan allows the subject to expect a particular form of update which is an incrementation.

Détienne observed that activation of elementary schemas such as variable schemas can allow the evocation of schemas representing algorithms that are partly composed of those elementary schemas. This is illustrated in the following rule:

IF SCHEMA1a (schema of variable): Counter_Variable plan: initialization I:= Z,
name I
IF SCHEMA1b (schema of loop): while A <> B
THEN SCHEMA1 (algorithmic schema): Linear_Search plan
THEN SCHEMA1a (schema of variable): Counter_Variable plan: update: I:= I+ 1

In this example, the initialization, the Counter variable and the while loop evoke an algorithmic schema for linear search. The elementary schemas that permit this activation are a Counter variable for which subjects infer the update and a schema of loop.

Other rules describe activation processes that are conceptually driven (the direction of activation is top-down); this is expressed as 'IF SCHEMA1 and..., THEN SCHEMA2' (SCHEMA 2 is a subcategory compared to SCHEMA1) or as 'IF SCHEMA1, THEN SCHEMA1a, THEN SCHEMA1c, ..., THEN SCHEMA1n'. Détienne (1988) gives an example of this process that allows the expectation of a complex combination of several schemas implemented in the code.

Norcio (1982) showed that semantic constructs can be evoked and inferences drawn from expectations based on comments in the program code. He asked subjects to fill in a blank line either at the beginning of a chunk or in the middle of a chunk. In one experimental condition, comments were inserted in the program at the beginning of each chunk. Results show a positive effect of comments compared to no comments in a fill-in-the-blank task when the line to fill in is at the beginning of a chunk. In that case, the comment provides the subjects with a cue for schema evocation.

Widowski and Eyferth (1986) have collected data on reading strategies used by programmers for programs varying along a dimension of stereotypeness. They remark that the strategies used by experts are different for usual and unusual programs. When reading usual programs, the activity seems to involve a conceptually driven processing. When reading unusual programs, they describe a more bottom-up oriented processing.

It should be noted too that in some cases, the evocation of schemas may create negative effects on performance inasmuch as the presence of inferred values is not confirmed in the code. Détienne (1984) observed this kind of negative effect in an experiment in which experts had to debug a program written by somebody else and where they had difficulties detecting errors when they strongly expected a particular value in the code (the actual value being different and incorrect). The experts failed to verify whether or not the value was actually present. This kind of negative effect is more likely to happen when the program has been written by the programmer, since this will create stronger expectations of what should be in the program.

In summary, data support the hypothesis that programming plans are evoked in program comprehension. The evocation of schemas allows inferences to be drawn. However, it may happen that there are no cues in the code to evoke schemas. Mental execution has been shown to be used to infer the goal of a part of code when the programmer has no expectation about it. By acquiring knowledge about the intermediate values of the variables during execution, the programmer can infer the goal of the process, and so, the goal of the part of code they have executed. Experts debugging programs have been observed to mentally simulate while not having enough textual cues about the goal of a part of code: no name, no familiar structure, no documentation (Détienne, 1984). This result suggests that schema evocation may be based either on static information like 'beacons' or 'dynamic information' which change with the program execution. This last kind of information is similar to the data-flow relations in a program.

Détienne and Soloway (1988) noted that mental simulation can also be used to infer information about the interactions between plans. In an experiment in which experts had to perform a fill-in-the-blank task on programs and had to verbalize while performing that task, they show that whatever the 'planliness' of programs is, the experts use simulation when they want to check for unforeseen interactions. This is typically used to understand programs in which a loop is used with a counter, suggesting that experts know by experience that the use of a count plan in a program can cause unforeseen interactions and that the best way to check for these interactions is to execute the part of the program with the count.

4.2.2 Inferences collected in recall tasks

Inferences collected in recall tasks subsequent to comprehension reflect the kind of knowledge that has been evoked in memory. Results of some studies support the hypothesis of semantic structures whereas other results support the hypothesis of syntactic structures.

In two experiments conducted by Détienne (1986b), experts were asked to recall a program following a debugging task. During the debugging phase, the subjects had to comprehend an unknown program and to evaluate its correctness. The programs were written in Pascal.

Distortions of the form of the program were observed in the recall protocols. For example, distortions concerned the name of a variable used as a counter. The subjects recalled I instead of J. This suggests that the variables are memorized in a category, a Counter_Variable plan, and that the lexical form is not kept in memory. In the recall process, the subjects use another possible value of the slot 'name of variable'.

Other observations suggest the existence of prototypical values in the slots of a schema. Each slot is associated with a set of possible values, as seen before. Those values have not the same status, some values being more representative of a slot for a schema than the others. When there is a prototypical value in a category, this value comes to mind first when the category is activated.

For example, in a schema for a flag, the slot 'context' is an iteration that can take the values 'repeat' or 'while'. In the program used, the variable, V, appears in a 'repeat' iteration. During the debugging phase, the value of the iteration expected by most of the subjects was 'while not V do'. In the recall protocols, a distortion was observed: a subject has recalled the instruction as 'while' instead of 'repeat'. This subject has reported the prototypical value instead of the adequate value. These observations give support to the hypothesis of prototypical values being associated to the slots of the schemas.

Distortions of the content of the program were also observed in the recall protocols. When a schema is activated, information associated to this schema is inferred. So information typical of a schema may be recalled while not included in the program text. Détienne reports a thematic insertion that was observed. For the particular domain of stock management, the following values could be assigned to the slots of the problem schema:

Task domain: stock management

Data structure: record(name of file, descriptor of file)...

Functions: allocation (creation or insertion), destruction, search

The function of creation was not isolated in a subprogram and the subjects did not read any information concerning this function in the program. Nevertheless, a subject has reported this function as if it was a subprogram. This suggests that schematic knowledge dependent on the task domain has participated in the elaboration of the representation and in the process of recovering of stored information at recall.

Thus, the data suggest that the knowledge structures evoked in comprehension are organized according to the semantics of that information. However, other data suggest that:

(1) experts also use knowledge structures that are organized in function of syntactic information (procedural);

(2) a representation based on this kind of knowledge may be constructed before one based on semantic knowledge (functional). This is compatible with the control-flow approach.

In Pennington's study (1987), subjects were asked to read a short program before answering questions. Results show that questions about control-flow relations are answered faster and more correctly than questions about data flow and functional relations. Furthermore, in a recognition memory test, subjects recognized a statement faster when it was preceded by another segment of the same control structure than when it was preceded by a segment of the same functional structure. This suggests that knowledge structures representing control flow have an important role in program understanding. Results also suggest that the understanding of the program control structures may precede the understanding of program functions.

However, results also suggest that understanding may be schema based or control based according to the understanding situation. Pennington remarks that there was an effect of the language. Two languages were used in the experiment: Fortran and Cobol. Cobol programmers were better at questions about data flow than were Fortran programmers, whereas control flow relations were less easily inferred by Cobol programmers. Pennington also remarks that there was an effect due to the comprehension task, i.e. the goal the subjects had while understanding the program. A modification task produced a shift toward increased comprehension of function and data flow at the expense of control-flow information.

4.2.3 Chunks constructed in program understanding

A chunk is the result of the identification of units belonging to a same mental construct. So a chunking task enables the study of the cognitive units on which comprehension is based. Several studies highlight the fact that the chunks constructed by experts and novices in understanding are different and that experts' chunks reflect semantic structures. This is compatible with the schema-based approach.

As programmers analyse programs on the basis of units that correspond to cognitive constructs, it is likely that highlighting those units in programs would facilitate the understanding process. Norcio (1982) has shown this type of effect by indenting programs on the basis of functional chunks defined by programmers and asking subjects to fill in a blank line either at the beginning of chunks or in the middle. The results indicate that subjects with indented programs supply significantly more correct statements compared to the non-indented group.

Black *et al.* (1986) and Curtis *et al.* (1984) remark that elements of code that are parts of the same plan are often dispersed in a program and that this characteristic of interleaving (see Chapter 1.2) is language dependent. This dispersion makes these parts difficult to integrate into a functional whole. This characteristic is part of what makes program comprehension hard. Letovsky and Soloway (1986) illustrate how difficult a program is to understand when using what they call a 'delocalized plan', i.e. a plan whose parts are dispersed in the program.

Rist (1986) provided experts and novices with programs to describe in terms of groups of lines of code that 'did the same thing'. In describing a program, novices used a mixture of syntactic and plan-based chunks. Experts used almost only plan-based groupings. Rist notices that when programs are complex, plan use decreases. In that case, construction of a mental representation of the program cannot be done from plan structure only, thus subjects use control-based program understanding. This last finding suggests that understanding may be schema based or control based according to the comprehension situation.

4.2.4 Comprehension mechanisms in different tasks

Comprehension is involved in different tasks of maintenance. Two tasks have been studied: enhancement (or modification) and debugging. In both of these tasks, studies show that experts may evoke programming plans so as to construct a representation of the program. However, studies also stress the importance of simulation mechanisms. This highlights the role of information on data-flow relations and control-flow relations in these tasks. This is compatible both with the schema-based approach and the control-flow approach.

In the task of modification, mental simulation has been observed to be used by experts. As simulation allows the inference of information on connections and interactions between functional components, it is likely to be used in tasks like enhancement in which processing this kind of information is particularly useful. Results from an experiment by Littman *et al.* (1986) show that some experienced programmers use this symbolic execution so as to acquire causal knowledge which permits them to reason about how the program's functional components interact during reading. Symbolic simulation is used to understand data flow and control flow.

This kind of strategy is the most effective inasmuch as it prevents subjects from introducing errors by modifying a part without taking into account the relationship between that part and other parts in the program. Simulation is a way to evaluate the external coherence between plans, i.e. to check whether or not there are interactions between plans and between goals (Détienne, 1989).

As reported before, Pennington (1987) noticed a marked shift toward increased comprehension of program function and data flow at the apparent expense of control-flow information following a modification task.

Simulation (i.e. mentally executing the program with values) is particularly useful in tasks like debugging in which it is important to judge whether the values produced at the execution are the expected values. In the debugging task, experts have been observed to mentally simulate the program so as to have information on the values taken by variables during the program execution (Détienne, 1984; Vessey, 1985).

It is noteworthy that mental simulation has also been observed in design tasks (see Chapter 3.3). It is not surprising, however, as the design task involves some comprehension. It is used so as to predict potential interactions between elements of the design and to check parts of the programs.

5 Discussion

Empirical data support the schema-based approach of program understanding. However, data also support the control-based approach. Studies show that according to the understanding situation, knowledge used may be related to different kinds of information: data-flow relations, functional relations or control-flow relations. Programming plans formalize information on data flow and functions whereas syntactic constructs reflect more the structure of the program as described with control-flow relations.

However, the two approaches of understanding are not contradictory. First, as said before, syntactic constructs may also be formalized as schemas. Secondly, a model of experts' knowledge may integrate those different kinds of knowledge. Thus, it should characterize the understanding situation so as to account for what kind of knowledge is used, when it is used and how it is used.

The understanding situation may be described by the characteristics of the language, the task, the environment and the subject. Concerning the language, the presence of cues in the code which allow the activation of schemas representing semantic knowledge may be dependent on the notational structure. Green (see Chapter 1.2) assumes that languages vary along a dimension of 'role-expressiveness'. With a 'role-expressive' language, the programmer can easily perceive the purpose, or role, of each program statement and, thus, the schemas evoked may be based on static cues whereas, with non-role-expressive languages, they may be based on the extraction of dynamic information like control-flow information.

Studies of programming have been developed on the basis of experiments conducted with a relatively narrow sample of programming languages: mostly Pascal, rarely Lisp, more rarely Basic, and more recently Prolog. It seems important now to conduct studies with other languages (for example, object-oriented languages) so as to take into account the effect of languages' characteristics in the understanding activity.

A model of the expert should be extended to take into account task variations. As Gilmore and Green (1984) remark, the information needed to be extracted from the code is different according to the task in which the programmer is involved. Furthermore, a particular language may make explicit in the code certain information or a particular environment may emphasize certain information important to achieve a particular task. So knowledge used by the expert may depend on the goal of his/her activity and on the availability of information in a particular situation.

References

Abelson, R. P. (1981). Psychological status of the script concept. *American Psychologist,* **36(7),** 715-729.

Adelson, B. (1981). Problem solving and the development of abstract categories in programming languages. *Memory and Cognition,* **9(4),** 422-433.

Atwood, M. E. and Ramsey, H. R. (1978). Cognitive structures in the comprehension and memory of computer programs: an investigation of computer program debugging. US Army Research Institute for the Behavioral and Social Sciences. Technical report (TR-78-A21), Va: Alexandra.

Black, J. B., Kay, D. S. and Soloway, E. (1986). Goal and plan knowledge representations: from stories to text editors and programs. *In* J.M. Carroll (Ed.), *Interfacing Thought: Cognitive Aspects of Human-Computer Interaction.* Cambridge, MA: MIT Press.

Bower, G. H., Black, J. B. and Turner, T. J. (1979). Scripts in memory for text. *Cognitive Psychology,* **11,** 177-220.

Brooks, R. (1983). Towards a Theory of the comprehension of computer programs. *International Journal of Man-Machine Studies,* **18,** 543-554.

Curtis, B., Forman. I., Brooks. R., Soloway. E. and Ehrlich. K. (1984). Psychological perspectives for software science. *Information Processing and Management,* **20(12),** 81-96.

Détienne, F. (1984). Analyse exploratoire de l'activité de compréhension de programmes informatiques. *Proceeding AFCET, séminaire 'Approches Quantitatives en Génie Logiciel'.* 7-8 June 1984, Sophia-Antipolis, France.

Détienne, F. (1986a). La compréhension de programmes informatiques par l'expert: un modèle en termes de schémas. Thèse de doctorat. Université Paris V. Sciences humaines, Sorbonne, 1986.

Détienne, F. (1986b). Program understanding and knowledge organization: the influence of acquired schemata. *Proceedings of the third European Conference on Cognitive Ergonomics*, Paris, September 15-20, 1986. (To appear in Falzon, P. (Ed). (1990). *Psychological Foundation of Human-Computer Interaction*. London: Academic Press.)

Détienne, F. (1988). Une Application de la Théorie des Schémas à la Compréhension de Programmes. *Le Travail Humain*, numéro spécial *Psychologie Ergonomique de la Programmation*, 1988, **51(4)**, 335-350.

Détienne, F. (1989). A schema-based model of program understanding. Eighth interdisciplinary workshop on 'Informatics and Psychology'. Schaärding (Austria), May 16-19 (to appear in *Mental Models in Human-Computer Interaction*. Amsterdam: North-Holland).

Détienne, F. and Soloway, S. (1988). An empirically-derived control structure for the process of program understanding. Research report 886, INRIA (to appear in International Journal of Man-Machine Studies).

Gilmore, D.J. and Green, T.R.G. (1984). Comprehension and recall of miniature programs. *International Journal of Man-Machine studies,* **21**, 31-48.

Letovsky, S. and Soloway, E. (1986). Delocalized plans and program comprehension. *IEEE Software,* **3(3)**, 41-49.

Linger, R.C., Mills, H. D. and Witt, B. I. (1979). *Structured Programming: Theory and Practice.* Reading, MA: Addison-Wesley.

Littman, D.C., Pinto, J., Letovsky, S. and Soloway, E. (1986). Mental models and software maintenance. *In* E. Soloway and S. Iyengar (Eds), *Empirical Studies of Programmers: First Workshop.* Norwood, NJ: Ablex.

Minsky, M. (1975). A framework for representing knowledge. *In* P. Winston (Ed.), *The Psychology of Computer Vision.* New York: McGraw-Hill. pp. 211-277.

Norcio, A.F. (1982). Indentation, documentation and program comprehension. *Human Factors in Computing Systems,* **15-17**, 118-120.

Pennington, N. (1987). Stimulus structures and mental representations in expert comprehension of computer programs. *Cognitive Psychology,* **19**, 295-341.

Rich, C. (1981). Inspection methods in programming. MIT AI Lab, Cambridge, MA., Technical report TR-604, 1981.

Rist, R. (1986). Plans in programming: definition, demonstration, and development. *In* E. Soloway and S. Iyengar (Eds), *Empirical Studies of Programmers: First Workshop.* Norwood, NJ: Ablex.

Rumelhart, D. E. (1981). Understanding understanding. Report, Center for Human Information Processing, University of California, San Diego, California.

Schank, R. and Abelson, R. (1977). *Scripts-Plans-Goals and Understanding.* Hillsdale, NJ: Lawrence Erlbaum Associates.

Shneiderman, B. (1976). Exploratory experiments in programmer behavior. *International Journal of Man-Machine Studies,* **5(2)**, 123-143.

Shneiderman, B. (1980). *Software Psychology.* Cambridge, MA: Winthrop.

Soloway, E. and Ehrlich, K. (1984). Empirical studies of programming knowledge. *IEEE Transactions on Software Engineering,* **10(5)**, 595-609.

Soloway, E., Ehrlich, K. and Bonar, J. (1982a). Tapping into Tacit Programming Knowledge. *Human Factors in Computer System,* **15-17**, 52-57.

Soloway, E., Ehrlich, K., Bonar, J. and Greenspan, J. (1982b). What do novices know about programming? *In* A. Badre and B. Shneiderman (Eds), *Directions in Human Computer Interaction.* Norwood, NJ: Ablex.

Vessey, I. (1985). Expertise in debugging computer programs: a process analysis. *International Journal of Man-Machine Studies,* **23**, 459-494.

Widowski, D. and Eyferth, K. (1986). Representation of computer programs in memory. *Proceeding of the Third European Conference on Cognitive Ergonomics.* Paris, September 15-19.

Chapter 3.2

Expert Programming Knowledge: A Strategic Approach

David J. Gilmore

Psychology Department, University of Nottingham, Nottingham NG7 2RD, UK

Abstract

This chapter considers an alternative to the 'programming plan' view of programming expertise, namely that expert programmers have a much wider repertoire of strategies available to them as they program than do novices. This analysis does not dispute that experts have 'plan-like' knowledge, but questions the importance of the knowledge itself versus an understanding of when or how to use that knowledge. This chapter will firstly examine evidence which challenges the completeness of the plan-based theory and then it will look at evidence which reveals explicit strategic differences. As the studies are presented a list of strategies available to experts will be maintained. Although the emphasis of this section of the book is on expert performance, the chapter concludes with a brief look at studies of novices, since the strategic approach includes the assumption that the strategies used by novices are different from those available to experts. From all these studies it will be seen that experts choose strategies in response to factors such as unfamiliar situations, differing task characteristics and different language requirements, whilst many problems for

Psychology of Programming
ISBN 0-12-350772-3

novice programmers stem not only from lack of knowledge, but also from the lack of an adequate strategy for coping with the programming problem.
The previous chapter has presented studies of expertise in computer programming that have concentrated on the content and structure of expert knowledge. The dominant concept has been the 'programming plan', which has been proposed as the experts' mental representation of programming knowledge, and which has been used in the development of programming tutors and environments for novice programmers.

This chapter examines those aspects of expertise that are not easily explained by plan-based theories. It is intended that the strategic approach described in this chapter should be seen as a complementary, rather than alternative, explanation of expertise. As the various studies are described a list of important strategies which programmers may use will be developed. The studies that are to be presented reveal problems with two implications of the plan-based approach:

(1) that 'programming plans' provide a complete explanation of expert programming behaviour;

(2) that a novice can be defined as someone who does not possess this expert knowledge.

An important, but often implicit assumption of the plan-based theorists is that the cognitive processes underlying programming are relatively straightforward. Often based on ideas from artificial intelligence, these processes are taken to be general problem-solving skills (cf. SOAR, Laird *et al.*, 1987; ACT* , Anderson, 1983), which a novice programmer also possesses. The learning of computer programming is the acquisition of the appropriate knowledge structures for the problem-solving skills to use. These knowledge structures may then be labelled 'plans' (see Chapter 3.1 for more details), though other possibilities exist. The critical feature is that expertise is seen as the acquisition of knowledge.

The alternative position is that expertise in programming may involve a variety of cognitive processes which, coupled with changes in knowledge, can give rise to a choice of different methods for solving any particular programming problem. These different methods can be termed strategies. The critical feature of the strategy argument is that observations of novice-expert differences can be caused by either knowledge differences, processing differences, or both.

Any assessment of this argument has two important strands. Firstly some evidence questioning the plan-based theory will be presented (evidence for the theory has been covered by the previous chapter) and then evidence of strategic differences will be described. For these reasons it is important that the evidence for alternative aspects to expertise beyond the plan-based theories is carefully considered

1 Generalizations of 'programming plan' theories beyond Pascal

The majority of research on programming plans has examined Pascal experts, since they are easily found in universities. However, if the theory is a truly accurate characterization of programming expertise the concept should generalize to other languages and paradigms.

```
program prob12;
vars depth, days, rainfall:integer;
     average:real;

begin
for days := 1 to 40 do
begin
depth := 0
writeln("Noah, please enter todays rainfall:");
readln(rainfall);
rainfall := rainfall + depth;
end;
average := depth / 40;
writeln("Average is", average);
end.
```

Figure 1: An example Pascal program with plans cued using different fonts (from Gilmore and Green, 1989). This program contains two errors.

One experiment that looked at the generalization issue was that by Gilmore and Green (1989), who examined the nature of programming plans in Pascal and Basic. They examined the ability of final-year undergraduate programmers (not expert, but comparable with groups used in similar research) to detect different types of errors, under different conditions of program presentation. The relevant condition for the current discussion was one in which the plan structure of the program was highlighted using different coloured highlighting pens. Figure 1 provides an example of this condition, using different fonts rather than different colours. The important result was that, for the Pascal programmers only, the highlighting led to a greater detection of plan-related errors (27% improvement) and to no improvement in the detection of other types of error (1% decrease). Thus, the Pascal programmers were responsive to plan structures, when performing plan-related tasks.

This result confirms previous work on the programming plans, though the task-specific nature of the effect is a departure from previous research. However, the results for the Basic programmers were quite different. In general they were not as responsive to structural cues as were the Pascal programmers, and they were least responsive to the plan-structure cues, which led to a 6% decrease in detection of plan-related errors and a 3% decrease in the detection of other errors.

Also important, given arguments that plans are the major aspect of expertise, was the result that Pascal programmers were also responsive to control-structure cues, but only when detecting control-flow errors (37% improvement, cf. 4% improvement on other errors). This indicates that according to the information available from the program the programmers were able to switch between different perspectives on the code.

From this study it became clear that the literal content of programming plans is not transferable between languages, even within the procedural paradigm. This means that we must either acquire a more precise theory of the development of programming plans, so that we can direct the differences across languages, or else we must perform new analyses of plan structures for every programming language. Gilmore and Green explained the effect in terms of difference in the *role-expressiveness* of Pascal and Basic.

Davies (1990) was unconvinced by this explanation of the language differences and repeated the experiments using a single language. He used the same methodology, but compared Basic programmers with and without formal instruction in software design skills. His results show that although programmers who had experience of structured programming techniques were responsive to plans (25% improvement in error detection), those programmers without experience of structured programming showed no such improvement (3%).

Since we have no information about the design skills of Gilmore and Green's subjects we cannot tell whether their experiment demonstrated the same effect as Davies and that what they interpreted as language differences were in fact due to differences in exposure to structured programming. Nevertheless, these results together pose a number of questions for advocates of plans as a theory of expertise in programming. These include 'where do plans come from?' and 'how are plans used in program generation?'.

The latter has been examined by Bellamy and Gilmore (1990) who collected protocols from a number of experts in different languages whilst generating a few simple programs. They attempted to compare Rist's model of plan use during coding (Rist, 1986) with the *'Parsing-Gnisrap'* model of Green *et al.* (1987). Unfortunately the results did not particularly support either model, since the evidence for plan use was poor. It seems that we are not yet at a stage where theories of the use of plans can be adequately described, since there is no usable, objective definition of a programming plan.

These three studies challenge the simplistic view of plans, but they do not give any cause to reject the concept completely. They provide a basis for arguing that much more research is needed on the psychological basis for plans, before we proceed with applying the theory. Furthermore, there is an increasing amount of research that reveals aspects of expertise which cannot be accounted for by theories of programming plans. It is to this literature that we now turn.

2 Alternative perspectives on expertise

The main problem with knowledge-based theories of expertise is that the learning process acquires knowledge about programming, not knowledge about how to do programming. Kolodner (1983) neatly captures this difficulty:

> even if a novice and an expert had the same semantic knowledge ...,
> the expert's experience would have allowed him to build up better episodic definitions of how to use it.

In fact, one of the early papers on programming plan theories of expertise (Chapter 3.1) included the learning of 'programming discourse rules', rules about conventions and styles in programming. Perhaps unfortunately, the emphasis since that paper has been almost solely on plans, leaving others to investigate episodic knowledge about how to do programming.

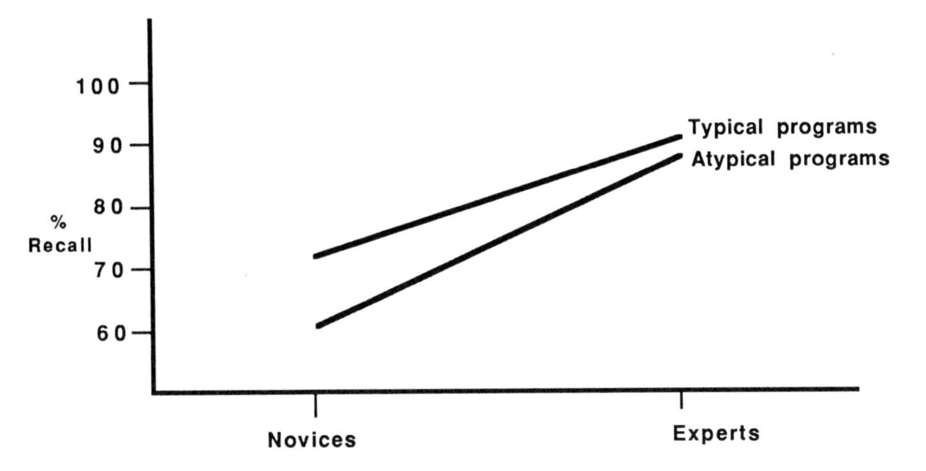

Figure 2: Widowski's (1987) results, showing that novices are, if anything, more affected by typicality than are experts.

3 Comprehension processes

The traditional application of the Chase and Simon (1973) chess research (see Chapter 1.4) is to compare novices and experts on well-ordered versus randomly ordered programs. But, since the effect is understood to be based on the expert's ability to readily extract meaningful chunks, the same paradigm can be applied to well-ordered programs that are more or less meaningful (typical or atypical). The knowledge-based theory would predict that novice-expert differences should be diminished with the atypical programs since neither the novices nor the experts will possess the plans to process it.

Widowski (1987) performed this comparison, varying both the degree of typicality (assessed through semantic complexity) and *structural complexity* (using the McCabe (1976) measure of syntactic complexity). He compared thirteen Pascal novices with thirteen Pascal experts in a programming environment that allowed them to view only one line of the program at a time. This enabled the collection of data about both the process and the results of comprehension. The subjects' task was to comprehend the program and then to reconstruct it from memory.

The recall results showed significant effects of expertise and of structural and semantic complexity. There was an interaction between the two forms of complexity, with the impairment due to semantic complexity being much greater in the structurally complex programs. But, the most striking result was the complete absence of an interaction between expertise and complexity. Contrary to the hypothesis, experts were better than novices for both typical and atypical programs. In fact, if anything the difference was greater for the atypical programs (see Figure 2).

On the data from the comprehension process there was evidence of plans guiding the experts' comprehension on the stereotypical programs, with a significant interaction between expertise and semantic complexity. On the atypical programs, experts seemed able to shift to a quite different strategy of comprehension. For novices the same strategy of comprehension seemed to be used for all programs. Structural complexity had no effect on comprehension processes.

In a second study, a similar task was used, but with verbal protocols collected during the comprehension process as well. In this the results of the first study were replicated. In the comprehension process two strategies were identified: control-structure oriented and variable-oriented. The main results were that experts varied the amount of structure-oriented processing according to the complexity of the program, but that they consistently used the variable-oriented strategy more than novices.

The identification of these two strategies, and the ability of experts to use either as needed, is consistent with the task-specific results of Gilmore and Green described above. Although Gilmore and Green did not explicitly address the comprehension process, it is possible to take all these results together, which suggests that the focus of a programmer's comprehension strategy is affected by the program presentation format (Gilmore and Green, 1989), the program's complexity (Widowski, 1987) and by the programmer's knowledge (Davies, 1990).

Whereas Widowski focused on the *process* of comprehension, Pennington (1987a) has looked at the mental representations that programmers form of a program. In her study she used a range of types of comprehension questions, each type accessing a different type of information from the program (e.g. control flow, data flow, function, state). She then performed a complex priming study, in which she recorded the time taken to recognize individual lines from the program. Her interest lay in whether the line to be recognized was immediately preceded by a line from the same control structure, or from the same plan structure. The hypothesis was that if programmers form plan-based mental representations, then they should recognize lines faster when preceded by lines from the same plan structure.

Her results were quite clear cut and contrary to the plan-based view of expertise. Her subjects made fewer errors on the control-flow questions, compared with the data-flow and function questions and the effect of priming was consistently greater when the prime came from the same control structure. These results were replicated in a second experiment, using a longer program. However, in this second experiment an extra stage was added, in which the programmers were asked to make a modification to the program, and half were asked to provide verbal protocols while doing so. Pennington found that after the modification phase the dominant mental representation was of data flow and function, especially for those who had supplied protocols. Thus, Pennington concludes:

> While plan knowledge may well be implicated in some phases of understanding and answering questions about programs, the relations embodied in the proposed plans do not appear to form the organising principles for memory structures.
> (Pennington, 1987a, p. 327).

Although these two studies did not lead to the identification of particular strategies, it seems reasonable to include task characteristics as another determinant of comprehension process (given that they change the nature of the mental representation of a program).

Pennington (1987b) investigated the differences between the best and worst programmers (cf. novice-expert comparisons). From an initial programming test she was able to identify the top and bottom quartiles of a group of professional programmers, who were then closely examined during a comprehension task. Comparing the

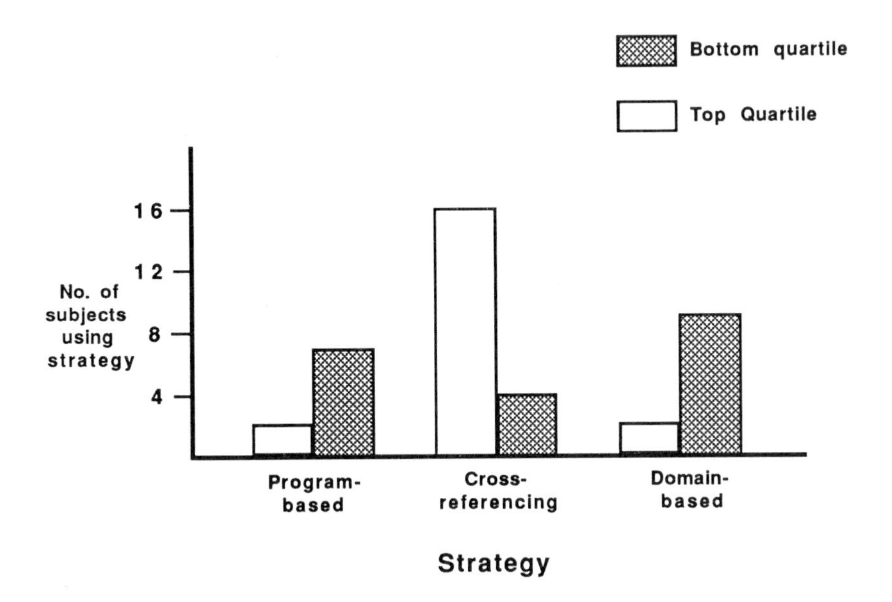

Figure 3: Pennington's (1987b) results revealing the different strategies used by her best and worst experts.

two groups of experts, she found that the most successful programmers adopted a *cross-referencing* strategy, considering not only the program, but also the real-world problem it addressed (Figure 3). This strategy can be contrasted with code-based and domain-based strategies, which were both commonly observed in the less successful experts.

In this result we can see that plan-based knowledge could be extremely important in the cross-referencing strategy, since plans (as described by Rist, 1986) provide bridges between domain knowledge and code fragments. However, it is quite possible that the less expert programmers possessed the same plan knowledge, but for some reason failed to use it.

From this review of studies about program comprehension, we can identify two important components of any strategy. Firstly there is the focus of attention, whether control flow, variables or maybe programming plans, and secondly, there is the scope of attention, which may be the code, the problem domain or both. Although not all of the combinations may actually occur in programming practice, the evidence suggests that a programmer's choice of strategy is influenced by his/her knowledge, the programming task, the program representation and the program's complexity.

4 Debugging strategies

A number of researchers have used the conventional novice-expert comparison for investigating debugging performance and some of these studies provide good examples of the need for a strategy approach. Using the implicit assumption that there is only one debugging strategy, researchers conclude that differences between expert and novice debuggers must be due to differences in their knowledge about either the program or programming in general.

Gugerty and Olson (1986) found that expert programmers were both more likely to find bugs, and less likely to introduce new bugs than were novice programmers, despite the fact that both groups engaged in similar comprehension activities prior to debugging. They conclude that these differences must be due to differences in the comprehension ability of the two groups since the same set of activities led to differing amounts of comprehension. However, there are alternative explanations. Firstly, the activities observed by the experimenters might have been similar manifestations of quite different comprehension strategies, in which case the debugging differences will follow from comprehension differences.

A more-likely difference is that similar comprehension strategies led to very different levels of understanding about the program, which then forced the two groups to use quite different strategies for debugging. Faced with inadequate comprehension the novices had little choice but to make changes to the code, in the hope that they would either be right, or lead to improved comprehension. Gugerty and Olson (1986) describe a number of strategies used by their subjects, including simulated execution of the program (or part of it), working backwards from the observed error in output, elimination of parts (through partial testing) and adding print statements.

Vessey (1985, 1987) looked at the debugging performance of a number of expert Cobol programmers. Her methodology was complex, involving various classifications of the expertise of her programmers, and advanced statistical analysis. An initial test established the chunking abilities of her programmers, using a variant of the standard recall paradigm (cf. McKeithen *et al.*, 1981), abilities that can be equated with plan knowledge. In a subsequent debugging task she classified the strategies of her subjects into two types: one based on erratic behaviour and reinspections of parts of the program already covered and the other based on a fluid appproach to problem solving and a smooth progression through the program. It was then possible, using analysis of covariance, to establish whether the strategy used by programmers when debugging a program was more important than the knowledge structures possessed prior to examining the program. Chunking ability was able to account for 31% of the variation in debugging time, whereas debugging strategy could account for 74%. Vessey concludes

> the chunks programmers possess may not be important factors in debugging expertise, i.e. programmers may possess effective debugging strategies that are more important to the debugging process than the programming knowledge they possess. (Vessey, 1987, p. 70).

Thus, the evidence from studies of debugging reveals that a number of debugging strategies are available and that the success of debugging is as much a property of strategy choice as it is one of knowledge structures. Unlike the discussion of comprehension strategies above, there is no direct evidence about those factors that influence choice of debugging strategy. However, it is fair to label experience (and expertise) as one such factor, given the number of studies that show marked novice-expert differences. Other factors may include program authorship (Waddington, 1989) and debugging environment (Gilmore, 1990).

5 Studies of novices

Although the emphasis of this whole section is on the nature of *expert* programming knowledge and performance, knowledge-based theories of expertise are often used to draw conclusions about teaching, which is regarded simply as the passing on of the expert's knowledge, and novice difficulties are assumed to derive from the lack of this knowledge. Thus, a brief part of our discussion about the need for a strategy or process-oriented approach to programming must be an examination of some of the studies of novice programming, which reveal that novices often display strategic, rather than knowledge-based difficulties.

Perkins and Martin (1986) conducted a series of interviews with students using Basic. They describe how the main difficulties were 'fragile knowledge' and 'neglected strategies'. The former is knowledge that the student has, but fails to use when it is needed. Evidence that the student possessed the knowledge is not simply the fact that they had been taught it, but that on nearly 50% of occasions where hints had to be given to students, the student went on to solve the problem, even though the hint did not contain the appropriate knowledge.

A component of the difficulty seems to be that the students were not reading the program to discover what it actually did ('close tracking' or 'parsing'). This is an example of a 'neglected strategy', a general problem-solving strategy that should be, but is not, used. Mistaken strategies also exist, as can be seen in one of their students who appeared to be obsessed with using syntactic features taught recently, even when the problem only required simple structures taught and mastered some weeks before.

Perkins and Martin (1986, p. 225) summarize their novices' problems as 'fragile knowledge exacerbated by a shortfall in elementary problem-solving strategies', suggesting by way of conclusion that we should not view programming as an opportunity for the development of general problem-solving skills, since they are themselves required for successful programming.

White (1988) conducted a study with Prolog programmers, some of whom had previously learnt Pascal. Although it is not surprising that interference effects occurred, what is interesting is that Prolog novices were able to use the appropriate terminology (knowledge) for the Prolog at the same time as using an inappropriate strategy from Pascal.

Bonar and Cunningham (1988), in describing their use of the intelligent tutoring system Bridge (developed out of plan-based theories), comment on how their students were quite successful at developing an outline solution using plan-based concepts, suggesting that they had acquired and understood the contents of the programming plans. However, the bottleneck for the students came when trying to translate the plan-based outline into actual Pascal code:

> Matching between the Phase 2 output and Pascal code was problematic, however. Because there is not always a simple match between a plan component and Pascal code, students will sometimes make a reasonable selection that Bridge doesn't accept. (Bonar and Cunningham, 1988, p. 409).

Thus, it seems that having knowledge of the plans alone is inadequate, it is understanding how to use them that counts.

Finally, in some unpublished studies of POP11 novices, I made some observations that support the idea that novices can know plans, but not be able to use them

```
define listdouble(list);
var result;
Omitted initial value for result
until list = [] do
  result <> [^ 2*hd(list))] -> result;
  tl(list) -> list;
enduntil;
pr(result);
enddefine;
```

(a) Iterative solution for simple list processing problem, with common error.

```
define listdouble(list);
if list = [] then
  Omitted terminal value for function
else
  [^ 2*hd(list))] <> listdouble(tl(list))
endif;
enddefine;
```

(b) Recursive solution for same list processing problem, with analogous error.

Figure 4: Iterative and recursive plans, as used by novice, with same error made repeatedly.

successfully. The students had been taught about both iteration and recursion and were regularly given problems requiring them to code both types of solution. The problems used were comparable, except that students could choose between iterative or recursive solutions. In a number of cases the student's response to an error message was to switch from the iterative to the recursive 'plan', or vice versa. In one remarkable protocol a student wrote an iterative solution (Figure 4a), but omitted a necessary initialization (the only error). Rather than try to edit the program, he simply switched to a recursive solution (Figure 4b), making the analogous error (failure to return a value from the stopping condition). He then switched back to the iterative plan, and so on back and forth a total of five times (ten attempts!), before the mistake was detected. The next problem was similar but a little more complex. He wrote an almost successful iterative solution, but again omitted the initialization. The only evidence of learning from the previous problem was that it only required five attempts to find this same (!) mistake.

Thus, it seems that possessing knowledge is not the only problem that novice programmers have. In a number of cases they show that they have the knowledge, but also that they do not know how to adequately use it. It seems that programming may be rather like riding a bike, or some other motor skill, in that without practice it cannot be mastered.

6 Conclusions

The simplest conclusion from this survey of the nature of programming skills is that expertise is not as simple as we might sometimes think. Although high-level, efficient representations of programming knowledge develop with experience, it seems that this knowledge is not the sole determinant of programming success. Besides the chunking of knowledge structures, experts seem to acquire a collection of strategies for performing programming tasks, and these may determine success more than does the programmer's available knowledge.

A number of strategies for use in comprehension and debugging tasks have been described, and it seems that these enable experts to respond more effectively to unfamiliar situations, differing task characteristics and different language requirements. Finally, and importantly for teachers of programming, a number of studies of novices' difficulties have shown that they derive not only from lack of knowledge, but also from lack of strategy.

References

Anderson, J. R. (1983). *The Architecture of Cognition.* Cambridge, MA: Harvard University Press.

Bellamy, R. K. E. and Gilmore, D. J. (1988). Programming plans: internal or external structures. *In* K. Gilhooly, M. Keane, R. Logie and G. Erdos (Eds), *Lines of Thinking: Reflections on the Psychology of Thought.* Chichester: Wiley.

Bonar, J. and Cunningham, M. (1988). *In* J. A. Self (Ed.), *Artificial Intelligence and Human Learning.* London: Chapman Hall.

Chase, W. G. and Simon, H. A. (1973). Perception in chess. *Cognitive Psychology,* **4,** 55-81.

Davies, S. (1990). The nature and development of programming plans. *International Journal of Man-Machine Studies,* **32,** 461-481

de Groot, A. D. (1965). *Thought and Choice in Chess.* The Hague: Mouton Press.

Gilmore, D. J. (1990). Models of debugging. Paper presented at *Psychology of Programming Interest Group: Second Workshop.* Walsall, January, 1990.

Gilmore, D. J. and Green, T. R. G. (1989). Programming plans and programming expertise. *Quarterly Journal of Experimental Psychology, HEP* **40(3),** 423-442.

Green, T. R. G., Bellamy, R. K. E. and Parker, J. (1987). Parsing-Gnisrap: A model of device use. *In* G. Olson, S. Sheppard and E. Soloway (Eds), *Empirical Studies of Programmers,* vol. 2. Hillsdale, NJ: Ablex.

Gugerty, L. and Olson, G. M. (1986). Comprehension differences in skilled and novice programmers. *In* E. Soloway and S. Iyengar (Eds), *Empirical Studies of Programmers.* Norwood, NJ: Ablex.

Kolodner, J. L. (1983). Towards an understanding of the role of experience in the evolution from novice to expert. *International Journal of Man-Machine Studies,* **19,** 497-518.

Laird, J. E., Newell, A. and Rosenbloom, P. S. (1987). SOAR: an architecture for general intelligence. *Artificial Intelligence*, **33**, 1-64.

McCabe, Th. J. (1976). A complexity measure. *IEEE Transactions on Software Engineering*, **2**, 308-320.

McKeithen, K. B., Reitman, J. S., Rueter, H. H. and Hirtle, S. C. (1981). Knowledge organisation and skill differences in computer programmers. *Cognitive Psychology*, **13**, 307-325.

Pennington, N. (1987a). Stimulus structures and mental representations in expert comprehension of computer programs. *Cognitive Psychology*, **19**, 295-341.

Pennington, N. (1987b). Comprehension strategies in programming. *In* G. Olson, S. Sheppard and E. Soloway (Eds), *Empirical Studies of Programmers*, vol. 2. Hillsdale, NJ: Ablex.

Perkins, D. N. and Martin, F. (1986). Fragile knowledge and neglected strategies in novice programmers. *In* E. Soloway and S. Iyengar (Eds), *Empirical Studies of Programmers*. Hillsdale, NJ: Ablex.

Rist, R. (1986). Programming plans: definition, demonstration and development. *In* E. Soloway and S. Iyengar (Eds), *Empirical Studies of Programmers*. Hillsdale, NJ: Ablex.

Shneiderman, B. (1976). Exploratory experiments in programmer behaviour. *International Journal of Computer and Information Science*, **5**, 123-143.

Vessey, I. (1985). Expertise in debugging computer programs: A process analysis. *International Journal of Man-Machine Studies*, **23**, 459-494.

Vessey, I. (1987). On matching programmers' chunks with program structures: An empirical investigation. *International Journal of Man-Machine Studies*, **27**, 65-89.

Waddington, R. (1989). Unpublished Ph.D. Thesis, University of Nottingham, 1989.

White, R. (1988). Effects of Pascal knowledge on novice Prolog programmers. *Proceedings of the International Conference on Thinking*. Aberdeen.

Widowski, D. (1987). Reading, comprehending and recalling computer programs as a function of expertise. *Proceedings of CERCLE Workshop on Complex Learning*. Grange-over-Sands.

Chapter 3.3

Expert Software Design Strategies

Willemien Visser[1] and Jean-Michel Hoc[2]

[1]*INRIA – Ergonomics Psychology Group, Rocquencourt BP 105, F-78153 Le Chesnay, France*
[2]*CNRS – Université de Paris 8, URA 1297: Psychologie Cognitive du Traitement de l'Information Symbolique, 2, Rue de la Liberté, F-93526 Saint-Denis, Cedex 2, France*

Abstract

Early studies on programming have neglected design strategies actually implemented by expert programmers. Recent studies observing designers in real(istic) situations show these strategies to be deviating from the top-down and breadth-first prescriptive model, and leading to an opportunistically organized design activity. The main components of these strategies are presented here. Consequences are drawn from the results for the specification of design support tools, as well as for programmers' training.

1 Introduction

Although the top-down and breadth-first design model is very elegant, and rather simple to describe and to understand, professional programmers are aware of the fact

Psychology of Programming
ISBN 0-12-350772-3

that it is difficult to implement it in real software design. When teachers of computer science prescribe this kind of design strategy, they know very well that they themselves cannot – and do not – implement it in their design practice. Teaching complex domain knowledge to novices surely requires simplifying the concepts and the procedures. Simple ideas can guide novices in developing more complex skills. We have to tell novices, however, that the difficulties they encounter in implementing these simple ideas are not only due to their lack of knowledge, but also inherently linked to the simplistic nature of these ideas. In addition, it would be a mistake to refuse the complexity of expert design strategies when developing design tools that, otherwise, would only be useful in the simple cases for which elegant strategies are implementable. Moreover, seriously taking into consideration the difficulties programmers have when implementing these 'optimal' strategies could orient tool designers towards defining tools that could support more efficient strategies.

This approach then requires the identification of the strategies implemented in real design activity. Relatively few studies in cognitive science handle the question of software design by experts. Most research in the domain of programming concerns other activities and other subjects: the activities examined are, in general, conducted on already existing programs (mostly comprehension and memorization, and to a lesser degree debugging) and the subjects studied are mostly novices (see Chapter 1.3). Of course these studies can make contributions to design models:

* Comprehension studies, on the one hand, show which representations (often formalized in terms of 'schemas', see Chapter 3.1) the programmer possesses. These structures are certainly used in design and they constrain the strategies which are implemented, but specific design studies have to show *how* they are used, and *under which conditions*. On the other hand, these studies may contribute to the understanding of comprehension activities involved in design, that is, the comprehension of problem specifications, of other designs, and of modules already written in the design in progress.

* Studies comparing novices to experts show differences that may be characteristic of expertise. Their results constrain the definition of objectives for learning programs and teaching methodologies, and the type of tools and assistance to be developed.

1.1 Design: resolution of ill-defined problems

The most common conception of design problems is considering them as 'ill-structured' ones, in contrast to 'well-structured' problems which the psychology of problem solving has almost exclusively studied until now (Eastman, 1969; Simon, 1973).

Often a task that constitutes a problem for the person who is solving it – that is, for which this person has a representation that cannot trigger a ready-made procedure to reach the goal (Hoc, 1988b) – is considered as 'ill defined' by this problem solver. Following the problem space formalism, the main feature of a design problem concerns its solution, that is, the goal to be reached (Hoc, 1988b). This goal (a program in software design, a blueprint in architectural design, etc.) is objectively ill-defined; if not (a program or a blueprint is available), the problem is solved and no design activity is any more required.

To reach a suitable goal representation, the designer uses and transforms intermediary representations which constitute rough, underspecified, or even inappropriate, approximations of the goal:

* generally, initial problem specifications are not sufficient to define the goal, and stepwise definition of new constraints is necessary;

* the resolution of conflicting constraints, often existing between different levels, plays an important role;

* specifications or constraints come from different representation and processing systems (see Chapter 1.3), are often conflicting and have to be translated into a specific design domain – in software design, this domain corresponds to the programming language used;

* there is no 'definite criterion for testing any proposed solution' (Simon, 1973, p. 183), such as there typically exists for 'well-structured' problems: design problem solutions are more or less 'acceptable' or 'satisfying', they are not either 'correct' or 'incorrect';

* various different solutions are acceptable, one being possibly more satisfying in one dimension, another in another dimension.

Software design, which concerns us in this chapter, exhibits these characteristics. For example :

* the specifications given at the start are never complete or without ambiguity (that is, in real work situations);

* different programs, implementing different algorithms, may do for solving one and the same problem, even if one program may be judged 'better' than another, often for reasons of execution or maintenance, that is, for machine- or user-oriented efficiency criteria.

According to the type of software design, certain characteristics are more or less strongly present: text-processing software is surely less constrained at the outset than a programmable controller program that is going to govern an automated process.

1.2 Early design studies: sticking to the prescriptions

At the beginning, a trend seen in design studies, but not only there, was to 'conflate prescriptive and descriptive remarks' on the activity, and, rather than to consider what the activity *is really like*, to focus on what it *should be* (Carroll and Rosson, 1985). This led authors especially to describe design as well structured and even as hierarchically organized. This same tendency is observed in most, but especially the early, software design studies, such as the one by Jeffries *et al.* (1981). In the introduction to their article, the authors describe existing design methodologies, arguing that:

* these methodologies are 'indicative of the guidelines that experts in the field propose to structure the task';

* 'a reasonable model of performance ... ought to be related to accepted standards of good practice';

* 'most expert designers are familiar with this literature and may incorporate facets of these methodologies into their designs' (p. 256).

As a matter of fact, these methodologies advocate (modular) decomposition and provide different bases for performing it. As will be seen later, however, studies on professional programmers show that the theoretically optimal methods whose use has been learned and recommended need, to be implemented in fact, very particular conditions, whose realization is often not assured.

2 Software design studies

2.1 Methodology used

2.1.1 Protocol studies

In most of the empirical studies focusing on software design, the data come from protocol analysis. A limited number of subjects (from one to ten) is asked to think aloud during their activity. Generally, the verbal protocols of only a subgroup of these subjects are analysed in detail.

Brooks (1977) was the first author to construct, from the protocols collected on one programmer writing a large number of short programs, a general model of the cognitive processes in computer programming.

2.1.2 More or less simple problem statements

Most of these studies (especially Jeffries *et al.*, 1981, and Adelson *et al.*, 1985) use experimenter-constructed, rather artificially limited problem statements.

Hoc (1988a) uses problem statements that lead to programs of small size, but which control algorithmic difficulties. These problem statements have been constructed in order to represent the categories of a typology of programming problems, resulting from a previous empirical study on problem classification in relation to the type of problem-solving strategy.

Guindon *et al.* (1987) judge the problem they use to be 'more complex and realistic ... than [those that have] been given in other studies of software design, yet not so different that [their] results cannot be easily compared to them' (p. 66). The problem they use is, however, incomparable, in complexity and degree of realism, with real design problems such as studied by Visser (1987).

Visser (1987) conducted a rather different, up to now – as far as we know – rather unique, study. Full-time observations were made on a professional programmer, in his real working environment (a machine tool factory), for a period of four weeks. The programmer was solving a real, complex industrial problem (control of a machine tool installation). Visser observed the programmer's normal daily activities without intervening in any way, other than by asking him to verbalize as much as possible his thoughts about what he was doing. In her data analysis, she focused on the specificity of the strategies used in real work conditions. Using the same method, observations were conducted on the mechanical engineer, during his design of the (functional) specifications for the programmer (see Visser, 1988a,b, 1990).

This methodological choice of observations on large software design projects is necessary. More often than not, observations or experiments are done on small software design problems and selection of problem statements is rather anecdotal. This leads to questioning the ecological validity of the results. Nevertheless, the size of the program is not the only relevant dimension for evaluating design difficulty. A large program, if it is familiar to the programmer, can take a long time to write, without any actual problem-solving activity, but requiring only coding. Conversely, a small program can accurately represent some problem-solving processes that are used in the development of large projects. As observations on small programs are more easily conducted on several programmers, they especially enable the evaluation of individual differences. Certain features of large software design, however, are impossible to reproduce when one reduces the program size (for example, the working memory management).

2.2 Models constructed

From his protocol analysis, Brooks (1977) elaborates a programming model, in which he distinguishes, next to the *coding* on which the model focuses, two other types of activities:

* *understanding*, leading to a problem representation;
* *method finding*, that is, construction of a plan, a (hierarchically) organized program representation (using relations between goals and subgoals) – an important role is occupied here by two kinds of knowledge structures, computer science and task domain-related schemas.

Adelson *et al.* (1985) propose a general model of the design activity, functioning with four components:

* a 'design meta-script', that is, a high-level schematic representation whose function is to drive the design process by setting goals for processing the 'sketchy model';
* the 'sketchy model', that is, the current solution state, which becomes progressively less sketchy, that is, more concrete and elaborate, until the implementation level representation;
* the 'current long-term memory set', consisting of all the known solutions appropriate to the aspect of the design that is currently being worked on;
* the 'demons', which monitor the state of the 'sketchy model', activating 'things to remember', provide elements to elaborate and modify the Sketchy Model into a final, concrete design solution.

Other authors stress the structures controlling the activity (Jeffries *et al.*, 1981; Guindon and Curtis, 1988).

Brooks (1977) and Jeffries *et al.* (1981) formulate production rules to account for the activity covered by the protocols. Adelson *et al.* (1985) use goals and operators organized in a goal hierarchy. Guindon and Curtis (1988) only describe the components of the model and their articulation.

3 Different strategies used in designing software

3.1 Variability between experts

Most authors note *variations in the design strategies* and the *solutions* they lead to, not only between levels of expertise when they compare experts and novices, but also between experts. Rather than considering this variability as a nuisance factor or a marginal result, we judge it as inherent and characteristic to design, as described in the introductory section, that is, the development of *a*, not *the*, solution to a problem that is not completely defined at the start. In such conditions, different designers will proceed in different manners, introducing different solution elements at different moments (see also Falzon and Visser, 1989, who show that, due to different past task experiences, different expert designers may exhibit different *types* of expertise).

3.2 Global control strategy: problem decomposition

Almost all subjects in early design studies are observed to use the same global control strategy advocated by design methodologies, that is, *decomposition* of the problem into subproblems. But several decompositions are possible for a problem, bearing on rather different principles.

In a study on more or less advanced computer science students, Ratcliff and Siddiqi (1985) identify two types of problem decomposition:

* data driven – the generation of the program structure is guided by a mental execution strategy, which bears on a simple representation of the input data, just sufficient to satisfy processing requirements;

* goal driven – the analysis of the goal structure leads to a non-trivial representation of the input data, which is more declarative than procedural, resulting in a quite different problem decomposition.

Two remarks may be formulated concerning the general nature of decomposition. Firstly, the use of this strategy may depend on expertise. The very beginner observed by Jeffries *et al.* (1981) does not succeed at all in decomposing problems, whereas their advanced students generate decompositions rather in terms of successive processing steps, than in terms of modules.

Secondly, as shown by Guindon *et al.* (1987), there are partial design solutions that are not the result of a decomposition, but which are, for example, retrieved 'by recognition'. As a consequence, such solution elements do not always fit into a balanced global solution.

If Jeffries *et al.* (1981), and Adelson *et al.* (1985) even more, stress the 'neat' predictable structure of decomposition strategies, and in general of the problem-solving activities involved in design, Guindon *et al.* (1987) and Visser (1988a,b) insist on the deviations of these structures. Moreover, the 'breakdowns' (Guindon and Curtis, 1988) observed do not always reflect deviations from predictable decomposition structures: many of them are rather caused by the opportunistic character of design.

3.3 Top-down and bottom-up strategies

The top-down strategy consists in descending the solution tree from the most abstract level down to the lowest, concrete level, never coming back up to a higher level (*top-down refinement*).

Based on his observations, Brooks (1977) expected that, only if a programmer is working on a problem with which he is very familiar and if he resolves it in a programming language in which he has considerable experience, he may proceed sequentially, without backtracking, through the three stages identified by the author (understanding, method-finding and coding). That is, only expert programmers, and even they only in these particular conditions, may proceed in a strictly top-down fashion.

Most design studies confirm this prediction.

For example, professional programmers working with a computer support to top-down processing that renders plan revision tedious generate non-optimal solutions: planning errors are sometimes rescued by awkwardly modifying modules at a very low level (Hoc, 1988a).

In the study of Jeffries *et al.* (1981), only one expert (out of four) showed a systematic implementation of a top-down strategy. The other experts deviated more or less from this 'optimal' strategy. The authors note that a designer may choose to deviate from the advocated order when he realizes that a component has a known solution, is critical for success, or presents special difficulties. This is typically one of the ways of proceeding qualified by Visser (1987) as 'opportunistic' (Section 3.5).

3.4 Breadth-first and depth-first strategies

Design methodologies describe and advocate the use of a breadth-first strategy – combined with a top-down strategy – for decomposition: when decomposing the current level solution, one should develop all the elements of the new solution at the same level of the solution 'tree' and integrate them into a new global structure, rather than refining, until its final solution elements, one or several particular solution branches (which would be *depth-first* processing).

All three experts observed by Adelson *et al.* (1985) implemented this strategy (called '*balanced development*' by the authors).

The one expert (out of four) in the study of Jeffries *et al.* (1981) observed to proceed systematically top down, did so in combination with a breadth-first strategy. Other experts were observed to follow a rather depth-first strategy, starting, for example, their decomposition by a top-down processing of only some branches of the tree, handling the other ones afterwards.

For handling interaction, breadth-first processing is of course very useful, even if the detection of potential interactions may require descending branches in anticipation.

Handling a problem at one level, one may think of related elements at another level. Sometimes, experts are capable of maintaining these kinds of elements in memory and retrieving them at the appropriate moment. Adelson *et al.* (1985) observe their experts making 'notes to themselves' (concerning constraints, partial solutions or potential inconsistencies) and the authors posit the existence of 'demons' reminding the designer to incorporate this information into the design once the appropriate level of analysis has been reached.

Hoc (1988a) and Guindon *et al.* (1987) observe, however, that even experts have difficulties considering and maintaining simultaneously all problem or possible solution elements at one level of abstraction. They observed subjects engaged in bottom-up processing activities and noticed the difficulties these subjects encountered, such as backtracking of subproblems whose solution had been postponed or whose solution had to be modified. In the situation analysed, Hoc ascribes these difficulties to the nature of the environment. Following a top-down strategy, the language and the editor used in this experiment constrain the subjects to express too precise expressions in the design. For example, when one is considering an iterative structure for a module at a high level in the tree, a precise representation of the adequate control structure can be unavailable. If the environment imposes the choice too early, the users have to analyse the problem at the next level. Then they write the result of this analysis in order to preserve their memory load and adopt a depth-first strategy.

In a study on expert programmers, Petre and Winder (1988) confirm the need for languages that can support different levels of analysis in the course of design. They notice that, before introducing the constraints of actual programming languages, experts very often use a personal pseudo-code.

3.5 Opportunistic strategies

As noticed above (see Section 3.3), only one expert (out of four) in the study of Jeffries *et al.* (1981) implements a 'pure' top-down strategy. The other three have been observed to display the following behaviours:

* starting the decomposition in the middle of the tree;
* working simultaneously on two distinct branches;
* making interruptions for digressions at other than the current level, for example, to deal with other subproblems or to define primitive operations, that is, elements at the lowest level;
* descending in the decomposition tree, but coming back up afterwards, for example to introduce a whole new solution decomposition level.

Guindon *et al.* (1987) noticed, at least, two types of returns: on the one hand, to tentative solutions proposed earlier. On the other hand, they observed that subjects, in an advanced design phase, re-examined the specifications to understand them. The designers studied by Visser (1988a,b) in their real working environment also proceeded to such re-examinations; however, they modified not only the specifications they received, but even those having governed anterior design stages and thus underlying their specifications.

At the onset of his design, the professional programmer observed by Visser (1987) decomposed the problem into functional modules which he planned to handle in a top-down way. Afterwards he often deviated from this plan, organizing his activity rather opportunistically (see Hayes-Roth and Hayes-Roth, 1979). Two important factors determining the guidance of his problem solving – causing the activity to be opportunistic – were the cognitive cost of actions and their importance (see Visser, 1990). Examples illustrating each one of these factors are the following:

On the one hand, if information required for handling the current design component was not, not yet, or not easily available, its processing was often *postponed* – because it would have been 'expensive' – leaving unfinished modules at the current level of the design. On the other hand, information was sometimes processed only because its processing was 'cheaper' than proceeding to the according-to-the-plan action. An example is the processing of the information provided by the information source at hand, rather than looking for the information required to handle the current design component. This then might lead to the programmer defining modules *in anticipation* and/or *at other levels of detail and abstraction* than the current one.

An action may be important because of the type of action or because of the object concerned by the action. For example, 'verifying' is an important action only if the verification concerns certain objects.

3.6 Prospective and retrospective, procedural and declarative strategies

The two types of problem decomposition identified by Ratcliff and Siddiqi (1985) were explained with reference to two design strategies: data driven or goal driven. After a study of problem classification by professional programmers in relation to design strategies, Hoc (1988b) proposed a more complex framework to classify problems and strategies from this point of view.

Data-driven strategies, as observed by Ratcliff and Siddiqi (1985), are of a *procedural* kind: the program is generated following a mental execution strategy. The statements are written in the order of execution and the writing is guided by an available procedure. This is a *prospective* procedural strategy very often encountered when the problem statements trigger a quite familiar procedure (for instance, execution by hand).

But a prospective strategy can be more *declarative*. This is the case, for example, in management problems where the complex structure of the input files and the relationships between them introduce strong and complex constraints on the program structure. Then a static representation guides the design. In other problems, the guide can be provided by the structure of the output files and their relationships: the program can be written following the reverse order and the strategy is *retrospective* declarative. These declarative strategies are quite well supported by diverse management programming methodologies (for example, Jackson or Warnier methods).

A retrospective strategy can be more procedural than declarative when guidance is given rather by goal-subgoal or preconditions relationships than by a static output file structure. This retrospective procedural strategy is often implemented by novices, as has been observed in physics problem solving (Larkin and Reif, 1979). The novices start from unknowns, they search for definitions of these variables in their data base, generate new unknowns (intermediary results), etc., until reaching given values. On the contrary, experts can classify problems in categories for which they have available procedures: they may then develop prospective procedural strategies.

In the experiment cited, Hoc (1988a) used this classification of problems and strategies to assess a programming environment especially designed to help professional programmers developing a structured, top-down and retrospective programming method. Although prospective strategies were not hindered by the environment, the experiment shows that retrospective strategies were strongly induced by the environment (whatever the type of problem). Prospective problems were solved with greater difficulties than retrospective ones. Nevertheless, retrospective problems were not so easy to solve in the environment: difficulties appeared to be explained by the lack of appropriate data structures in the language. When following a retrospective strategy, guidance cannot be given by the mental execution of the procedure and must be provided by data structures and relationships between them.

Results of this kind show that sticking to a too rigid design methodology, supposed to be valid whatever the type of problem (or subproblem), is to be avoided. The domain of validity of a methodology should rather explicitly be defined in terms of a typology of problems.

3.7 Simulation

Adelson *et al.* (1985) stress the frequent occurrence of this strategy and its importance for the experts they observed. The mental models their subjects constructed were run as they were elaborated at different levels of abstraction. These simulation runs are supposed to assist the designer on, at least, two points:

* predicting potential interactions between elements of the design;
* pointing out elements of the solution state that need expansion.

Maintaining the balanced development is considered by the authors to serve these simulations, which require all elements of the model that is run to be at the same level of detail.

Simulation may serve different objectives:

* comprehension, when the designer explores and simulates the problem environment;
* evaluation, when he runs simulations of tentative solutions, and selects between them, for example, on criteria of efficiency.

Guindon *et al.* (1987) observed that these 'exploratory' design strategies are used for (sub)problems for which specialized design schemas cannot be evoked. They insist especially on the high frequency of occurrence of the simulation-for-comprehension strategy in their study compared to the results obtained in other design studies.

The programmer observed by Visser (1987) used simulation of the program's execution, mainly to understand the specifications. Simulation of the installation's operation, generally considered to characterize novices, was sometimes used by this experienced programmer to check modules of the part of the program he had already written. These simulations were among the rare moments the programmer verbalized his thoughts spontaneously.

3.8 Reduction of complexity

Considering a problem under its most typical form, modifying it only later to take into account its specific conditions, is a cognitive economical strategy. It requires, however, that the designer has at his disposal categories of problems and associated solutions ('schemas'), differentiating these different elements by their appropriate attributes. So it seems to be reserved to expert designers.

Various authors observed this kind of strategy. So did Hoc (1988a), but he noticed the environment precluding the implementation of such a generalization strategy. Indeed, this approach made it difficult to modify an initially developed solution with the editor used in his experiment.

The designers observed by Guindon *et al.* (1987) used another type of complexity reduction strategy, when they generate simplifying assumptions which they afterwards evaluate for their plausibility.

3.9 Considering users of the system to be designed

Both Adelson *et al.* (1985) and Visser (1987) observe their subjects to be guided, more or less, by such considerations, but with different functions in the two studies.

In the first one, mental *simulation* of a user's interaction with the system helped the designer to *think of elements to be included* in the design.

For the programmer observed by Visser, the ease of use for future users (system operators as well as maintenance personnel) was a *criterion of evaluation* of his design: considering homogeneity an important factor of ease of use, this led him to *make the program as homogeneous as possible*.

One may suppose that this strategic consideration has not been observed to be implemented in more studies, because of its link to – and perhaps dependency on – real work situations. Adelson *et al.* (1985) noted it, however, even if – contrary to Visser (1987) – they made their observations in a restricted laboratory setting.

3.10 Use of past experience and other knowledge

The use of knowledge has been examined much more in program comprehension studies (see Chapter 3.1) than in the design studies presented here.

3.10.1 Use of software design knowledge

Solutions, next to being constructed in an actual problem-solving activity, may be arrived at by retrieval – and of course adaptation – of a stored solution, which may be modified or not. The retrieved solution may come:

* from memory – for example, algorithms learned by using them in the past, or a published algorithm that has been retained (Jeffries *et al.*, 1981);

* from an external source – for example, an earlier written program (Jeffries *et al.*, 1981).

Visser (1987) observed the programmer relying heavily on existing (partial) solution instantiations (that is, listings of programs written in the past and parts of the program-in-progress). Once again, one may think that most psychological studies paying so little attention to this reuse – contrary to software engineering, which considers it to be a major problem to be solved – is due to their rather artificially limited context making reuse difficult to implement: to reuse a design module in the resolution of a problem, this problem must be similar to those the designer has processed in the past.

3.10.2 Use of problem domain knowledge

Knowledge about exemplars of the kind of system to be designed or about functional requirements of systems in general may be used to constrain the definition of the design-in-progress or to retrieve solution elements (Adelson *et al.*, 1985).

Visser (1987) noted that schema-guided information processing could explain certain errors made by the programmer, for example, when he violated the specifications for an 'atypical' function, by defining it as a completely typical function. The programmer's expectations – based on prototypical schema slot values – were probably so strong that he did not take into account the values which were given (in the specification document) (see Détienne, 1990).

4 Assistance to the design activity

The results of the presented studies provide a rationale for the specification of assistance tools to support designers.

4.1 Displays for helping the management of working memory

One may think of tools enabling *parallel presentation of intra- or inter-level information* (see the difficulties on breadth-first decomposition) or presentation of all *constraints on the solution order* or *maintaining a trace of postponed subproblems* needing backtracking (see the 'Design Journal' suggested by Guindon and Curtis, 1988, which could have still other functions).

4.2 Libraries of design schemas

Visser (1987) noticed the importance of examples and of past designs reuse for the programmer she observed. As long as a designer wishes to use his own past productions, such libraries are not too difficult to realize. But the interest of this function lies especially in providing designers with the experience of colleagues. In this case, problems of indexing, for example, arise.

4.3 Assistance for the articulation of top-down and bottom-up components

As a purely top-down decomposition is rarely implemented, such assistance would be useful. In the environment evaluated by Hoc (1988a), the articulation between these two components was really precluded.

4.4 Assistance to prospective strategies

As Hoc (1988a) concludes, this kind of assistance is more difficult to implement than assistance to retrospective strategies (as provided in the environment he evaluated). Subjects' goals are indeed more difficult to infer than the prerequisites of explicitly enunciated goals.

Structure-based editors have been proposed in order to aid prospective strategies.

4.5 Assistance to simulation

Given the importance of simulation in progressively developed design, assistance to this function would also be useful. As simulation often involves holding simultaneously several variables in mind, one could think of supporting the management of this memory load – a function that would not be particular to, but especially important for, simulation assistance.

The development of all these tools requires more research into the processes and components of the strategies to be assisted.

5 Conclusion

The different studies presented in this chapter did not all come up with the same results. Especially between those of Jeffries *et al.* (1981) and Adelson *et al.* (1985), on one side, and those of Guindon *et al.* (1987) and Visser (1987), on the other side; an important difference concerns the systematic use of rather 'optimal' decomposition strategies, such as top-down and breadth-first processing, which the first authors noticed to characterize their designers. Guindon *et al.* (1987) and Visser (1987) seemed to observe many more deviations from these strategies, leading them to consider 'opportunism' as an important factor of design activity.

An explanation for this difference might be found in the type of design problem to be solved: real design problems (as used by Visser, 1987, and approximated by Guindon *et al.*, 1987) or restricted problems (as used by the others). Further studies of expert design activity on more or less realistic problems could confirm this hypothesis. Working on a computational geometry algorithm design, Kant (1985) notices that 'control [of design activity] ... comes out of responding to the data and out of the problems and opportunities arising during execution' (p. 1366). Ullman *et al.* (1987) conclude that 'mechanical designers progress from systematic to opportunistic behaviour as the design evolves' (p. 157).

However, the difference we notice may also be due (to some extent) to the perspective the authors take on their data. As mentioned in the introductory section, early design studies often stuck strongly to the normative viewpoint. Both studies concluding on the systematic nature of top-down and breadth-first strategies (Jeffries *et al.*,1981; Adelson *et al.*, 1985) are among the first studies conducted on expert software design. Psychologists know the role expectations play on the processing of information.

References

Adelson, B., Littman, D., Ehrlich, K., Black, J. and Soloway, E. (1985). Novice-expert differences in software design. *In* B. Shackel (Ed.), *Human-Computer Interaction – INTERACT 84*. Amsterdam: North-Holland.

Brooks, R. (1977). Towards a theory of the cognitive processes in computer programming. *International Journal of Man-Machine Studies*, **9**, 737-751.

Carroll, J. M. and Rosson, M. B. (1985). Usability specifications as a tool in iterative development. *In* H. R. Hartson (Ed.), *Advances in Human-Computer Interaction*, vol. 1. Norwood, NJ: Ablex.

Détienne, F. (1990). Program understanding and knowledge organization: the influence of acquired schemata. *In* P. Falzon (Ed.), *Cognitive Ergonomics: Learning and Designing HCI.* London: Academic Press, pp. 245-256.

Eastman, C.M. (1969). Cognitive processes and ill-defined problems: a case study from design. *In* D. E. Walker and L. M. Norton (Eds), *Proceedings of the First Joint International Conference on Artificial Intelligence.* Bedford, MA: MITRE.

Falzon, P. and Visser, W. (1989). Variations in expertise: implications for the design of assistance systems. *In* G. Salvendy and M. Smith (Eds), *Designing and Using Human-Computer Interfaces and Knowledge Based Systems.* Amsterdam: Elsevier.

Guindon, R. and Curtis, B. (1988). Control of cognitive processes during software design: What tools are needed? *In* E. Soloway, D. Frye and S. S. Sheppard (Eds), *CHI'88 Conference Proceedings.* Reading, MA: Addison-Wesley.

Guindon, R., Krasner, H. and Curtis, B. (1987). Breakdowns and processes during the early activities of software design by professionals. *In* G. Olson, S. Sheppard and E. Soloway (Eds), *Empirical Studies of Programmers: Second Workshop.* Norwood, NJ: Ablex.

Hayes-Roth, B. and Hayes-Roth, F. (1979). A cognitive model of planning. *Cognitive Science,* **3**, 275-310.

Hoc, J. -M. (1988a). Towards effective computer aids to planning in computer programming. Theoretical concern and empirical evidence drawn from assessment of a prototype. *In* G. C. van der Veer, T. R. G. Green, J. M. Hoc and D. Murray (Eds), *Working with Computers: Theory Versus Outcomes.* London: Academic Press.

Hoc, J.M. (1988b). *Cognitive Psychology of Planning.* London: Academic Press.

Jeffries, R., Turner, A. A., Polson, P. G. and Atwood, M. E. (1981). The processes involved in designing software. *In* J.R. Anderson (Ed.), *Cognitive Skills and Their Acquisition.* Hillsdale, NJ: Erlbaum.

Kant, E. (1985). Understanding and automating algorithm design. *IEEE Transactions on Software Engineering,* **11**, 1361-1374.

Larkin, J. H. and Reif, F. (1979). Understanding and teaching problem solving in physics. *European Journal of Science Education,* **1**, 191-203.

Petre, M. and Winder, R. (1988). Issues governing the suitability of programming languages for programming tasks. Paper presented at *ECCE4 - Fourth European Conference on Cognitive Ergonomics.* Cambridge, September 1988.

Ratcliff, B. and Siddiqi, J.I.A. (1985). An empirical investigation into problem decomposition strategies used in program design. *International Journal of Man-Machine Studies,* **22**, 77-90.

Simon, H. A. (1973). The structure of ill-structured problems. *Artificial Intelligence,* **4**, 181-201.

Ullman, D., Staufer, L. A. and Dietterich, T. G. (1987). Preliminary results of an experimental study of the mechanical design process. *Proceedings of the Workshop on the Study of the Design Process.* Oakland, CA, February 1987.

Visser, W. (1987). Strategies in programming programmable controllers: a field study on a professional programmer. *In* G. Olson, S. Sheppard and E. Soloway (Eds), *Empirical Studies of Programmers: Second Workshop.* Norwood, NJ: Ablex.

Visser, W. (1988a). Giving up a hierarchical plan in a design activity. Research Report No. 814. Rocquencourt: INRIA.

Visser, W. (1988b). Towards modelling the activity of design: an observational study on a specification stage. *Proceedings of the IFAC/IFIP/IEA/IFORS Conference Man-Machine Systems. Analysis, Design and Evaluation,* vol. 1. Oulu, Finland, June 1988.

Visser, W. (1990). More or less following a plan during design: opportunistic deviations in specification. *International Journal of Man-Machine Studies* (special issue on empirical studies of programmers), in press.

Part 4

Broader Issues

Chapter 4.1

The Psychology of Programming in the Large: Team and Organizational Behaviour

Bill Curtis[1] **and Diane Walz**[2]

[1]*Microelectronics and Computer Technology Corporation (MCC), 9430 Research Boulevard, Austin, TX 78759, USA*
[2]*Assistant Professor, Department of Accounting and Information Systems, The University of Texas at San Antonio, San Antonio, TX 78285, USA*

In its century-old quest to explain human behaviour, psychology has spawned many subfields to study different phenomena. Several of these psychological specialties, such as social, organizational, ecological and interactional psychology, focus on how humans behave in groups and how situational conditions affect them. Disappointingly little theory and few research paradigms from these fields have been imported into the psychological study of programming. Most of the empirical research on software development has been performed on individual programming activities (Curtis, 1985; Curtis *et al.*, 1986). This orientation toward programming as individual activity occurred because:

(1) most psychologists studying programmers were not social or organizational pychologists;

(2) experiments on team and organizational factors are difficult and expensive to conduct.

Psychology of Programming
ISBN 0-12-350772-3

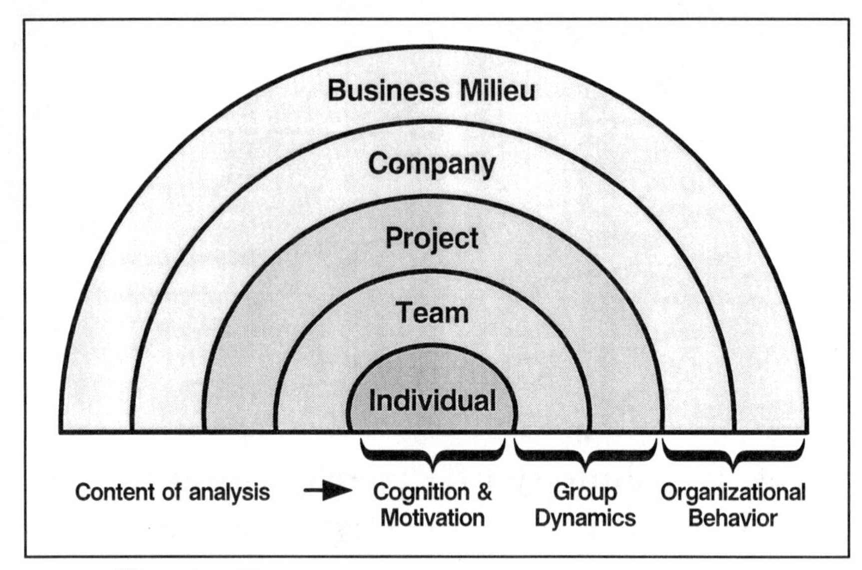

Figure 1: The layered behavioural model of software development.

Since software tools and practices are usually designed for individual problem solving, their benefits often do not scale up on large projects to overcome the impact of design processes that emerge from the social behaviour of the development team or the organizational behaviour of the company. Software development must be studied at several behavioural levels, as indicated in the *layered behavioural model* in Figure 1 proposed by Curtis *et al.* (1988). This model emphasizes factors that affect not merely the cognitive processes of software development, but also the social and organizational processes.

The layered behavioural model is an abstraction for organizing the behavioural analysis of large software projects. This model is orthogonal to traditional process models by presenting a cross-section of the behaviour on a project during any selected development phase. The layered behavioural model encourages researchers to extend their evaluation of software engineering practices beyond individuals, to determine if their effects scale up to an impact on the performance of an entire project (Kling and Scacchi, 1982; Scacchi, 1984). In order to develop a psychology relevant to programming in the large, we must take the *systems* view of software development activities encouraged by this model.

Contrary to most software process models, the layered behavioural model focuses on the behaviour of the humans creating the artifact, rather than on the evolutionary behaviour of the artifact through its developmental stages. At the individual level, software development is analysed as an intellectual task subject to the effects of cognitive and motivational processes. When the development task exceeds the capacity of a single software engineer, a team is convened and social processes interact with cognitive and motivational processes in performing technical work. In larger projects, several teams must integrate their work on different parts of the system, and *inter*team group dynamics are added on top of *intra*team group dynamics.

Projects must be aligned with company goals and are affected by corporate politics, culture and procedures. Thus, a project's behaviour must be interpreted within the context of its corporate environment. Interaction with other corporations either as co-contractors or as customers introduces external influences from the business milieu. The size and structure of the project determine how much influence each layer has on the development process.

Curtis (1988) described five psychological paradigms that have been used in studying programming. In the following sections we will review contributions to the empirical study of software development from two of these paradigms: *group dynamics* and *organizational behaviour*. Unfortunately, there is little research guided by these paradigms compared to the literature on the cognitive aspects of programming. Each of these two paradigms will be explored in later sections of this chapter. Although the senior author has conducted controlled experiments on programming phenomena (cf. Curtis *et al.*, 1989), the study of team and organizational behaviour has required different empirical methods, some more characteristic of sociology or anthropology. In subsequent sections we will describe some of the recent exploratory studies at MCC of software development projects from which we have tried to identify important problems that should become the focus of future empirical research on programming in the large.

1 Group dynamics

1.1 Programming team structure

Organizing programmers into teams superimposes a layer of social behaviour on the cognitive requirements of programming tasks. Two structures have been proposed for programming teams based on the centralized versus decentralized team organizations often studied in group dynamics research.

Mills (1971) and Baker (1972) designed a centralized organization for programming teams that placed primary responsibility for programming on a *chief programmer*. Other team members such as backup programmers, the program librarian and technical writers were organized as a support team for the chief programmer. Technical communication was *centralized* through the chief programmer. The success of this approach is generally believed to depend on the availability of a stellar technician to take the role of chief programmer.

In contrast to the chief programmer model, Weinberg (1971) proposed a decentralized team structure. In Weinberg's *egoless* team no central authority is posited in any team member. Different members take leadership responsibility for those project tasks that match their unique skills. The communication network in this team structure is *decentralized*, with technical information flowing freely among all team members. The key to egoless teams is that no single individual feels private ownership of any piece of the program. The program is a shared work product and decisions concerning it are reached by consensus. Weinberg recommended that the maximum size for this team was about ten members.

Social psychologists have established several results about centralized versus decentralized team structures relevant to differences between chief programmer and egoless teams. Shaw (1971) concluded that empirical evidence generally supported the following principles of team behaviour:

Table 1: Favourable conditions for different team structures.

Condition	Chief programmer	Egoless
Difficulty of problem	simple	complex
Size of program	large	small
Creativity required	low	high
Reliability requirements	low	high
Modularity requirements	high	low
Schedule	tight	relaxed
Duration of project	short	long
Team morale	low	high
Risk taking	low	high

* Groups usually produce more and better solutions to problems than individuals working alone, and their judgements are usually better on tasks involving error.

* A decentralized communication network is more effective for solving complex problems, whereas a centralized network is better for solving simple ones.

* Leaders emerge more often and organizational development is more rapid in centralized teams.

* A centralized network is more vulnerable to the saturation of its communication channels than is a decentralized network.

* Greater conformity and higher morale occur in decentralized teams.

Mantei (1981) used these and other principles to suggest conditions under which chief programmer and egoless teams would be most effective. An augmentation of her analysis is presented in Table 1. For instance, chief programmer teams should be most effective on large, simple, tightly schedule projects of short duration which do not require highly reliable or creative solutions. Egoless teams, on the other hand, should be most effective on small, complex projects requiring highly reliable and creative solutions with some risk performed under relaxed schedules over a long duration. Rather than being an either-or choice, chief programmer and egoless teams appear suited for different types of programming projects.

Unfortunately, few programming projects fit neatly into one of the two categories described above. For instance, many aerospace projects can be described as large, complex, tightly scheduled efforts spread over a long duration with high reliability requirements and sections requiring creative solutions. However, in such projects large portions of the system are not complex and do not require creativity. These latter components have been developed by the organization on previous projects and the structure of their solution is familiar to the project team.

A hybrid approach to structuring programming teams might be taken on large projects that have characteristics favourable to different types of programming teams.

Portions of the system whose solution does not present a new technical challenge might be programmed by chief programmer teams. Critical path or innovative portions covered by stringent reliability requirements or requiring creative solutions would be programmed by egoless teams. Within a single project, tasks would be assigned to the type of team best suited for them, thus matching the structure of the team and the task (von Mayrhauser, 1984). This matching is characteristic of the *interactional* approach that Sells (1963, 1966) has argued is necessary in making recommendations for real-world activities based on psychological theory.

In simulating team programming performance, Scott and Simmons (1975) found that as the amount of communication increased among the members of a five-person team, the chief programmer team became less productive because of the communication bottleneck which developed around the chief programmer. When programming team membership was varied between three and eighteen people, productivity leveled off between nine and twelve people, a level consistent with Weinberg's (1971) recommendation for the size of an egoless team.

1.2 The interaction of methodology and team process

Basili and Reiter (1981) were among the first to study actual programming teams experimentally. They wanted to determine the effects of programming discipline on team performance. Their experiment involved forty-five advanced students assigned to one of three conditions. These conditions were seven three-person teams trained in a disciplined team methodology, six three-person teams provided with no methodological training, and six individuals working alone. The task was to develop a compiler for a small, high-level language that was estimated to require two person-months and 1200 lines of Simpl-T.

Basili and Reiter used automated data collection techniques to amass measures on both the product and the process. They found almost no differences between the three development approaches on the product measures at traditional ($p < 0.05$) levels of statistical significance. However, differences among the approaches emerged on the process measures. The disciplined teams required fewer computer job steps, fewer compilations, fewer executions, and fewer changes than the unorganized teams or individuals.

Based on these results, Basili and Reiter concluded 'that methodological discipline is a key influence on the general efficiency of the software development process'. They believed there was evidence, although weaker, 'that mental cohesiveness is a direct influence on the general quality of the software development product...and that the disciplined methodology offsets the mental burden of organizational overhead [of working in teams] and enables a disciplined programming team to behave more like an indiviudal programmer relative to the developed software product'.

Boehm *et al.* (1984) studied seven teams of programmers using either a prototyping or a top-down specification-based methodology for producing a moderate sized application. Although the prototyped application contained less code and required less development effort, the traditionally specified packages exhibited a more coherent design and were easier to integrate. The prototyped packages were easier to learn and use, but were less functional and robust. These two approaches appeared to focus on different attributes of the problem during design. Prototyping focused on the user, while traditional specification-based approaches focused on the structural integrity of the program. Deciding which approach is more appropriate depends on an analysis of various trade-offs during the design process.

Some programming team research has studied technical reviews. Myers (1978) found that team walkthroughs consumed twice as many minutes per defect found as did individual execution testing. However, walkthroughs exhibited much less variability in results, since the large differences among individuals yielded large differences in the effectiveness of individual testing sessions. In a further analysis, Myers found that pooling the results of independent testing sessions was more effective than team walkthroughs in detecting defects. He observed differences in the focus of these different testing approaches. Individuals focused too much on normal rather than abnormal conditions, while walkthrough teams focused too much on logic rather than input/output problems. A fruitful area of future research concerns how to use a mix of individual and team processes to focus attention on different aspects of a problem.

Programming team activities offer many opportunities for peer review activities that may be formal or informal components of the development process. Shneiderman (1980) reported several brief experiments in having programmers provide anonymous feedback on the design of each other's programs. Most programmers indicated that the experience was educational, although it was naive to expect that anonymity could be maintained when programmers knew each others' coding habits from previous exposure. Similar educational benefits for team walkthroughs were observed by Lemos (1979).

1.3 The MCC object server study

Although existing studies help us begin to build a psychology of programming team behaviour, we need deeper insight into the actual processes that occur as teams develop programs over extended periods of time. The existing research on programming teams either studied the quantitative results of a task completed in a short time, or assessed the output of an extended programming assignment performed by students. In order to study the behaviour of an actual programming team, we collected longitudinal data on an MCC team that was building an object server to provide a repository for the persistent objects created during object-oriented programming. For a period of three months, we videotaped every group meeting held among customers, developers, and combinations of both groups. We transcribed seventy-two hours of videotapes covering thirty-seven different meetings that ranged from one to two hours.

The analysis of these videotapes emphasized the project's information requirements and their effect on the group process, especially its information sharing activities (Walz *et al.*, 1987). We began by assuming that the conflicts within a software design team were not solely the result of incompatible goals and/or opinions, but also represented the natural dialectic through which knowledge was exchanged. A scheme was developed for coding each utterance by a team member into one of several categories that reflected the unique characteristics of software design meetings. Analyses of these utterances indicated that the design meetings were generally dominated by a few individuals on the team to whom fellow participants attributed the greatest breadth of expertise. These individuals appeared to form a coalition that controlled the direction of the team.

An important issue in group dynamics that has not been discussed in the context of programming teams is the formation of coalitions. A small subset of the design

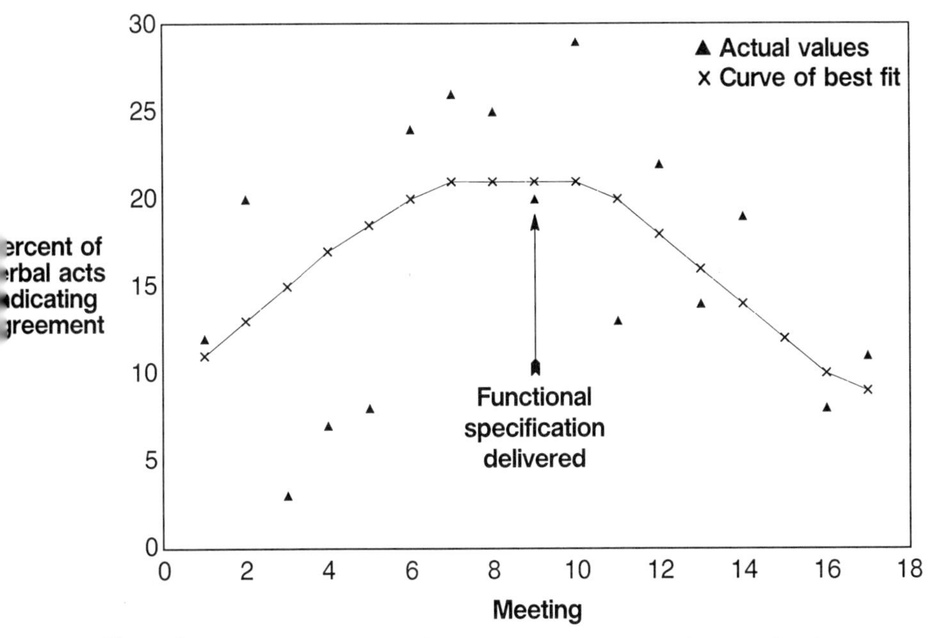

Figure 2: Level of agreement among designers working together on subtasks.

team with superior application domain knowledge often exerts a large impact on the design. Dailey (1978) found that collaborative problem solving was related to productivity in small, rather than large, research and development teams. Similarly, the small, but influential, design coalitions we have seen in other design projects represent the formation inside the larger team of a small team within which collaboration was more effective. Exceptional designers were often at the heart of these coalitions. Although teams usually outperform the average team member, the phenomena of exceptional designers is important because there is little evidence that teams can outperform the best team member (Hastie, 1987; Kernaghan and Cooke, 1986).

A scheme was developed to categorize each verbal act in the meeting protocols according to its role in the meeting (Walz, 1988). An interesting pattern in the verbal acts representing agreement among participants was observed across the seventeen meetings in the design phase of this project. A simplistic model assuming that co-operative design activity requires agreement among team members would hypothesize that such agreement should increase monotonically across meetings. However, as the data in Figure 2 demonstrate, there was a surprising inverted U-shaped curve (verified through logistic regression) that characterized verbal acts of agreement. This plot presents the percentage of agreement among team members working on the same subtask; a conservative analysis that should produce more consistent and lasting patterns of agreement than analyses based on the entire team. However, the plot indicates that agreement increased until the period (meetings 7 to 10) when the design team released a document presenting its response to customer requirements. In subsequent meetings the level of agreement began to decrease.

Since we analysed data from only a single team, it is difficult to draw conclusions that we would generalize to other software design episodes. However, there are some intriguing possibilities that lead to radically new models of how design teams operate. For instance, we may have observed a phenomena in which a team must reach some level of agreement in order to produce an artifact, but then reopens conflicts that were suspended in order to meet a milestone. On the other hand, we may have observed a phenomena where reaching consensus at one level of abstraction in a top-down design process does not relieve the conflicts that will result from issues that emerge at the next level down. Alternatively, we may have observed the impact of a small coalition that took control of the team during a period of deadline pressure and forced the level of consensus necessary to produce an artifact, but then dissolved when the deadline pressure passed, opening the door to renewed disagreement. Finally, we may have been observing a phenomena isolated to the unique characteristics of this particular team. We will not be able to make a definitive assessment of conflict resolution in design teams until we have studied other teams and compared our data to those produced by other researchers who study design teams longitudinally.

The design behaviours we observed in the meetings we videotaped suggested a three-stage process. During the first stage the team focused on determining requirements through learning about the application. The second stage involved communicating requirements among team members and with customers in order to develop a common understanding of the system to be built. In the third stage the group focused on creating artifacts that satisfied the requirements. Although the first and third stages are typical in most models of the software development process, the second stage involving a dialectic to surface misunderstandings is rarely made explicit.

Our observational study raised important questions that we need to study further in order to improve software development methods and process models. Consensus decisions are best achieved when a team reaches common understandings about their disagreements on interpreting requirements and how different architectural models of the system might operate. Our observations suggest the importance of explicit stages for training in the application domain and surfacing assumptions about the design. In these stages group conflict may be an important precursor to establishing group consensus. Formal requirements analysis methods may need process components that indicate when to enter and exit these various stages.

1.4 Team behaviour summary

Unfortunately, there has been too little research on software development teams in relation to their impact on software productivity and quality. If we assume that the conceptual unity of the program design is critical to the success of the software project, then teams must co-ordinate their work to make it appear as the work of one individual. The advantage of teams is their ability to bring divergent perspectives to bear in designing a program architecture. The process of synthesizing diverging design opinions on a programming team into a consensus involves conflict. Far from being a destructive force within the team, conflict can be a creative force if managed properly. Team methodologies must focus on co-ordinating the tasks and product concept. The structure of programming teams should: (1) reflect the nature of the task rather than the organization, (2) allow members to speak as if with one mind, and (3) determine the tasks the team can effectively handle.

2 Behaviour in programming organizations

2.1 The impact of organizational factors on programming

Although the primary focus for increasing software productivity and quality has been on improved methods, tools and environments, their impact in empirical studies of software productivity and quality on large projects from several industrial environments has been disappointing compared to the impact of factors characterizing the behaviour of the organization. For instance:

* In IBM Federal Division, Walston and Felix (1977) found that the complexity of the customer interface, the users' involvement with requirements definition, and the experience of the project team had more productivity impact than the use of software methods and tools.

* In the Defense and Space Group at TRW, Boehm (1981) found that the capability of the team assigned to the project had twice the productivity impact of the complexity of the product, and four times the impact of software tools and practices.

* In identifying a broad set of factors that acounted for two-thirds of the variation in software productivity, Vosburgh *et al.* (1984) argued that half of this variation was affected by factors over which project management had little control (factors other than the use of software engineering practices and tools).

In these studies human and organizational factors presented boundary conditions that limited the situations in which methods, tools and environments could increase software productivity and quality.

Many of the technologies purported to improve software productivity and quality only affect a few of the factors that exert the most influence over outcomes on large projects (Brooks, 1987). As the size of the system increases, the social organization required to produce it grew in complexity and the factors controlling productivity and quality may change in their relative impact. Technologies that enhance productivity on medium-sized systems may have less influence on large projects where factors that, while benign on medium systems, ravage project performance when unleashed by a gargantuan system that may involve the co-ordination of many companies. A great danger on large projects is that management will be deceived by the simplicity of the prescribed processes and will not understand what pitfalls are likely to await them (Fox, 1982).

Most of the organizational research on computing organizations has been on the effect of computing systems on the structure and functioning of the organizations that use the systems. Far less research has been performed on how organizational behaviour affects those who produce the systems. In part, this is because the cost of conducting experimental studies on these whole divisions is prohibitive. One of the few topics investigated has been the power and influence of information services departments in large organizations. Lucas (1984) reported several reasons that information services departments are perceived to have lower power than other departments in a company and carry less influence in decision making than might be expected. Srinivasan and Kaiser (1987) warn about the level of exposure that the programming team is allowed from outside sources. However, these studies have given us little insight into the problems that corporations experience in designing large, complex computer systems.

2.2 The MCC field study

In order to obtain insight into large system development problems, members of the software empirical research team at MCC conducted a field study of large software development projects (Curtis *et al.*, 1988). The field study was designed to provide detailed descriptions of development problems in such processes as problem formulation, requirements definition and analysis, and software architectural design. We sought to study projects that involved at least ten people, were past the design phase, and involved real-time, distributed or embedded applications. We interviewed projects from nine companies in such businesses as computer manufacturing, telecommuncations, consumer electronics and aerospace. Our objective was to get software development personnel to describe the organizational conditions that affected their work. From their descriptions we hoped to gain insight that could lead to better models of actual development processes at several levels of analysis, especially the team and organizational levels of the layered behavioural model presented in Figure 1.

In this study we employed *field research* methods characteristic of sociology and anthropology (Bouchard, 1976). This field study consisted of interviews with ninety-seven project team members from seventeen large system development projects. We conducted hour-long structured interviews on-site with systems engineers, senior software designers, the project manager, and occasionally the division general manager, customer representatives, or the testing/QA manager. Participants were guaranteed anonymity, and the information reported was 'sanitized' so that no individual person, project, or company could be identified. The data we collected lend themselves to the creation of the case studies that Benbasat *et al.* (1987) and Swanson and Beath (1988) recommended for use in research on information systems development and project management.

Three of the most important problems we uncovered in this study were the thin spread of application domain knowledge on the software development staff, fluctuating and conflicting requirements, and communication and co-ordination breakdowns. We will describe one of the problems we investigated – breakdowns in communication and co-ordination – as a fertile source of opportunities for psychological research on programming in the large at the organizational level.

2.3 Communication and co-ordination breakdowns

One of the major problems in organizational behaviour that we identified in the field study was breakdowns in project communication and co-ordination. A large number of groups had to co-ordinate their activities, or at least share information, during software development. Figure 3 presents some of the groups mentioned during field study interviews, clustered into behavioural layers according to their remoteness from communication with individual software engineers (cf. Tushman, 1977). Remoteness involved the number of nodes in the formal communication channel that information must pass through in order to link the two sources. The more nodes that information had to traverse before communication was established, the less likely communication was to occur.

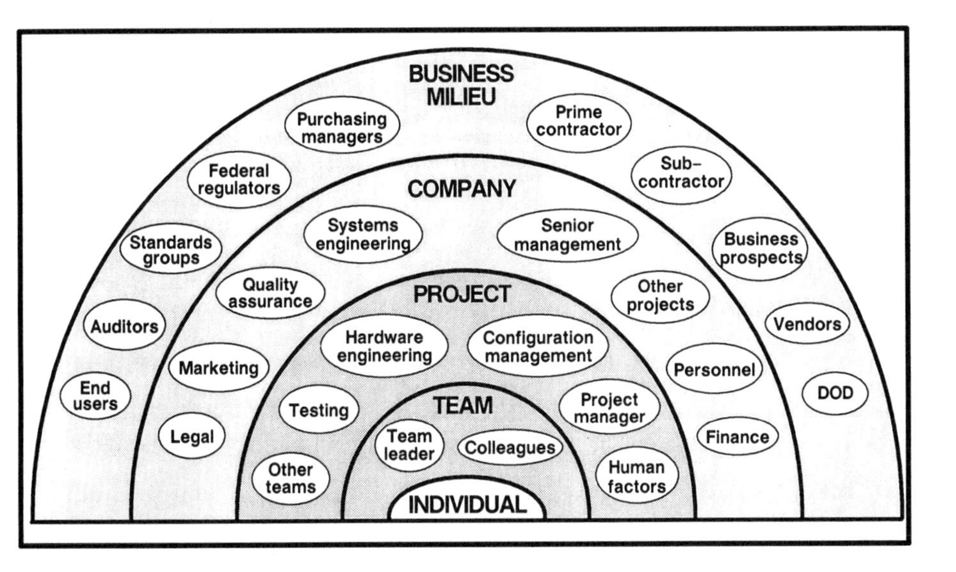

Figure 3: Remoteness of communications expressed in the layered behavioural model.

The model in Figure 3 implies that a software engineer normally communicated most frequently with team members, slightly less frequently with other teams on the project, much less often with corporate groups, and except for rare cases, very infrequently with external groups. Communication channels cross these levels were often preconditioned to filter some messages (e.g. messages about the difficulty of making changes) and to alter the interpretation of others (e.g. messages about the actual needs of users). In addition to the hindrances from the formal communication structure, communication difficulties were also due to the geographic separation, to cultural differences, and to environmental factors.

Organizational boundaries to communication among groups both within companies and in the business milieu inhibited the integration of knowledge about the system. These communication barriers were often ignored since the artifacts produced by one group (e.g. marketing) were assumed to convey all the information needed by the next group (e.g. system design). However, designers complained that constant verbal communication was needed between customer, requirements and engineering groups. For instance, organizational structures that separated engineering groups (hardware, software and systems) often inhibited timely communication about application functionality in one direction, and feedback about implementation problems that resulted from system design in the other direction.

Most project members had several networks of people they talked with to gather information on issues affecting their work. Similar to communication structures observed by Allen (1970) in research and development laboratories, each network might involve different sets of people and cross organizational boundaries. Each network supported a different flow of information; for example, information about the application domain, the system architecture, and so forth. When used effectively, these sources helped co-ordinate dependencies among project members and supplemented their knowledgte, thus reduced learning time. Thus, integrating information from these different sources was crucial to the performance of individual project members (Allen, 1986). These networks lend themselves to the network analysis described by Rogers and Kincaid (1981).

Some communication breakdowns between project teams were avoided when one or more project members spanned team or organizational boundaries (Adams, 1976). One type of *boundary spanner* was the chief system engineer, who translated customer needs into terms understood by software developers. Boundary spanners translated information from a form used by one team into a form that could be used by other teams. Boundary spanners had good communication skills and a willingness to engage in constant face-to-face interaction; they often became hubs for the information networks that assisted a project's technical integration. In addition, they were often crucial in keeping communication channels open between rival implementation teams.

On most large projects, the customer interface was also an *organizational communications* issue and this interface too often restricted opportunities for developers to talk with end users. The formal chain of communication between the developer and the end user was often remote because the communication chain traversed through the programmer's team leader, the team leader's project manager, the project manager's marketing organization, marketing's contact with the purchasing manager, and ultimately the end user for which the system was being purchased (Figure 4). Notice that four nodes have been placed between the programmer and the end user.

This tortuous chain of communication links was usually set up to solve two communication problems. First, the marketing group wanted to control the customer's access to the developers so that any attitudes or personal habits of the programming staff that would shake the customer's confidence in the development organization are not observed. Secondly, the marketing group wanted to establish a single point of contact for the customer, so that consistent messages would be sent to the customer through a single source. However, at the same time the interface was often cluttered with communications from non-user groups such as auditors, finance and standards groups in the customer's organization, each with its particular concerns. Typically, development organizations could not get a single point of customer contact for defining system requirements. Since no single group served as the sole source of requirements in either commercial or government environments, organizational communications became crucial to managing the project.

Designers needed operational scenarios of system use to understand the application's behaviour and its environment. Unfortunately, these scenarios were too seldom passed from the customer to the developer. Customers often generated many such scenarios in determining their requirements, but did not record them and abstracted them out of the requirements document. Lacking good scenarios of intended system use, designers worked from the obvious scenarios of application use and were unable to envision problematic exception conditions or subtleties the customer would ultimately want. In some cases, classified documents contained operational scenario information, but the software designers could not obtain the documents from the customer, because developers were not considered to have a need to know. There

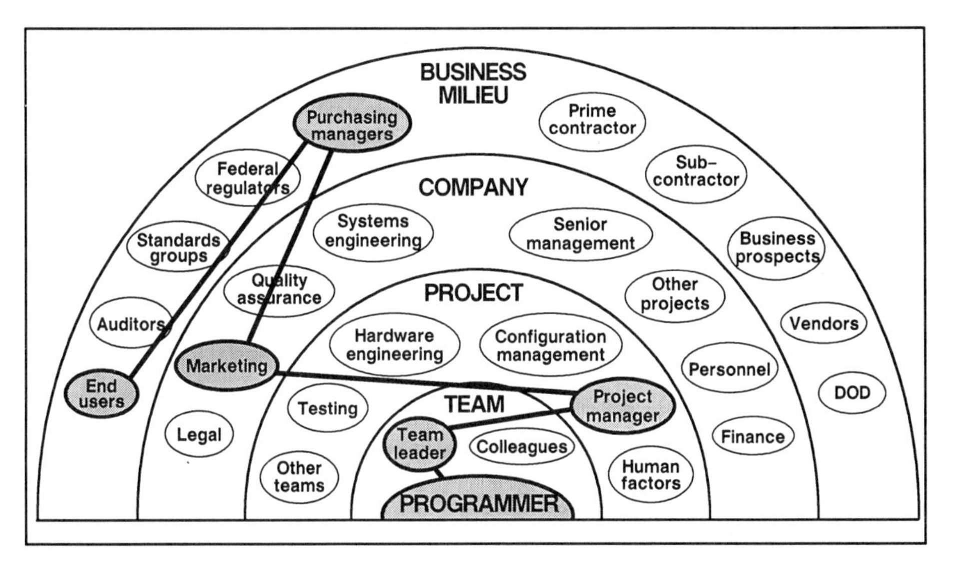

Figure 4: The long path from the programmer to the end user.

might have been less need for system prototypes meant to collect customer reactions if information generated by potential users had been made available to the project team. That is, many projects spent tremendous time rediscovering information that, in many cases, had already been generated by customers, but not transmitted.

Large projects required extensive communication that was not reduced by documentation. When groups such as marketing, systems engineering, software development, quality assurance and maintenance reported to different chains of command, they often failed to share enough information. This problem is not surprising in a government environment where security requires tactical information about a system to be classified. However, even on commercial projects information was occasionally withheld from a development group for reasons ranging from political advantage to product security.

Project staff found the dialectic process crucial for clarifying issues. Particularly during early project phases, teams spent considerable time defining terms, co-ordinating representational conventions, and creating channels for the flow of information. Artificial, often political, barriers to communication among project teams created a need for individuals to span team boundaries and to create informal communication networks. The complexity of the customer interface hindered the establishment of stable requirements, and increased the communication and negotiation costs of the project.

2.4 Software development questions for organizational psychologists

The paradigm of organizational behaviour provides multiple levels for analysing external impacts on programming teams. These levels include the parent organization, the local division, the customers, the business marketplace, the management structure, the administrative procedures, the physical environment, the psychological environment, the professional environment, etc. Although there is a large body of empirical research on organizational behaviour (Dunnette, 1976), little of it has been performed in programming organizations. Most current thinking on organizational processes in software development comes from accounts in books like those by Weinberg (1971), Brooks (1975), Kidder (1981), Fox (1982) and DeMarco and Lister (1987). This area badly needs new research, especially with the growing size and complexity of software products, and their effect on the size and complexity of the organizations that build them. Some of the most important questions concerning how organizational processes affect software development are:

* What is the most effective way to structure a programming organization, tall (many layers of management) or flat (large span of management control)?

* Does matrix management provide greater control over programming projects or does it reduce motivation by denying programmers a sense of ownership and pride in the software product?

* Under what conditions does a technical career path which parallels that of the management path help retain top programming talent?

* What are the major factors affecting the morale and climate of programming organizations?

* Which programming functions (e.g. testing, quality assurance, documentation, etc.) should be managed in groups separated organizationally from the software development group?

* How should the physical environment be arranged to maximize performance?

* Under what conditions will flex hours, quality circles, and other quality-of-work-life techniques be effective in software oganizations?

* How should development organizations interact with their customers, especially when the customer is a complex organization like the US Department of Defense?

Introductions to these issues can be found in many books on organizational behaviour and industrial psychology, and the results of current research can be perused in such journals as *Administrative Science Quarterly*, the *Journal of Occupational Psychology*, the *Journal of Applied Psychology*, *Personnel Psychology*, *Organizational Behavior and Human Decision Processes*, *Human Relations*, the *Journal of Management*, the *Academy of Management Journal* and the *IEEE Transactions on Engineering Management*.

3 Conclusion

Programming in the large is, in part, a learning, negotiation, and communication process. These processes have only rarely been the focus of psychological research on programming. The fact that this field is usually referred to as the 'psychology of programming' rather than the 'psychology of software development' reflects its primary orientation to the coding phenomena that constitute rarely more than 15% (Jones, 1986) of a large project's effort. As a result, less empirical data has been collected on the team and organizational aspects of software development.

Although there are cognitive questions involved in programming in the large (Curtis *et al.*, 1988), they must be investigated with an understanding of the social and organizational processes involved. For instance, given the amount of knowledge to be integrated in designing a large software system, and given the inability of current technology to automate this integration (Rich and Waters, 1988), the talent available to a project is frequently the most significant determinant of its productivity (Boehm, 1981; McGarry, 1982). But contributions by good people come not just from their ability to design and implement programs. Good people must become involved in myriad social and organizational processes such as resolving conflicting requirements, negotiating with the customer, ensuring that the development staff shares a consistent understanding of the design, and providing communications between two contending groups.

The constant need to share and integrate information on a software project suggests that just having smart people is not enough. The communication needed to develop a shared vision of the system's structure and function, and the co-ordination needed to support dependencies and manage changes on large system projects are team and organizational issues. Individual talent operates within the framework of these larger social and organizational processes. The influence of exceptional designers is exercised through their impact on other project members, and through their ability to create a shared vision to organize the team's work (Brooks, 1975). Recruiting and training must be coupled with *team building* (Thamhain and Wilemon, 1987) to translate individual talent into project success. Thus, the impact of processes at one level of the layered behavioural model presented in Figure 1 must be interpreted by their impact on processes at other levels.

The requirements levied on advanced software systems are growing rapidly (Boehm, 1987). Larger organizations – often webs of organizations – are required for producing systems that can require well over 10 000 000 lines of code. Companies often bet their future in a particular market on their ability to create a large software system rapidly. For instance, in 1986 ITT Corporation sold off their traditional telecommunications business because of difficulties in developing System 12, a fully distributed digital switch. The issues of how to organize programmers into teams and how to manage large programming organizations are crucial to many companies' success. Unfortunately most of the information available for guiding decisions about social and organizational structure in software organizations is anecdotal or is an extrapolation of results found with student teams in laboratories. The scientific community performing empirical research on programming must approach software development from more than a cognitive paradigm, and should begin performing research on the social and organizational aspects. At the same time corporations must make their software development organizations available for research, so that they are not forced to discount results obtained on undergraduate student teams. Better research on team and organizational factors may increase our ability to account for variation in software productivity and quality, and thus our ability to manage large systems development.

References

Adams, J. S. (1976). The structure and dynamics of behavior in organizational boundary roles. *In* M. D. Dunnette (Ed.), *Handbook of Industrial and Organization Psychology.* Chicago: Rand-McNally, pp. 1175-1199.

Allen, T. J. (1970). Communication networks in R&D laboratories. *R&D Management,* **1(1)**, 14-21.

Allen, T. J. (1986). Organizational structure, information technology, and R&D productivity. *IEEE Transactions on Engineering Management,* **33(4)**, 212-217.

Baker, F. T. (1972). Chief programmer team management of production programming. *IBM Systems Journal,* **11(1)**, 56-73.

Basili, V. R. and Reiter, R. W. (1981). A controlled experiment quantitatively comparing software development approaches. *IEEE Transactions on Software Engineering,* **7(3)**, 299-320.

Benbasat, I., Goldstein, D. K. and Meand, M. (1987). The case research strategy in studies of information systems. *MIS Quarterly,* **11(3)**, 369-386.

Boehm, B. W. (1981). *Software Engineering Economics.* Englewood Cliffs, NJ: Prentice-Hall.

Boehm, B. W., Gray, T. E. and Seewaldt, T. (1984). Prototyping versus specifying: a multiproject experiment. *IEEE Transactions on Software Engineering,* **10(3)**, 290-302.

Boehm, B. W. (1987). Improving software productivity. *IEEE Computer,* **20(9)**, 43-57.

Bouchard, T. J. (1976). Field research methods. *In* M. D. Dunnette (Ed.), *Handbook of Industrial and Organization Psychology.* Chicago: Rand-McNally, pp. 363-413.

Brooks, F. P. (1975). *The Mythical Man-Month.* Reading, MA: Addison-Wesley.

Brooks, F. P. (1987). No silver bullet. *IEEE Computer,* **20(4)**, 10-19.

Curtis, B. (Ed)., (1985). *Human Factors in Software Development,* 2nd edn. Washington, DC: IEEE Computer Society.

Curtis, B. (1988). Five paradigms in the psychology of programming. *In* M. Helander (Ed.), *Handbook of Human-Computer Interaction.* Amsterdam: Elsevier North-Holland, pp. 87-105.

Curtis, B., Soloway, E., Brooks, R., Black, J., Ehrlich, K., and Ramsey, H. R. (1986). Software psychology: the need for an interdisciplinary program. *Proceedings of the IEEE,* **74(8)**, 1092-1106.

Curtis, B., Krasner, H. and Iscoe, N. (1988). A field study of the software design process for large systems. *Communications of the ACM,* **31(11)**, 1268-1287.

Curtis, B., Sheppard, S. B., Kruesi-Bailey, E., Bailey, J. and Boehm-Davis, D. (1989). Experimental evaluation of software documentation formats. *Journal of Systems and Software,* **9(1)**, 1-41.

Dailey, R. C. (1978). The role of team and task characteristics in R&D team collaborative problem solving and productivity. *Management Science,* **24(15)**, 1579-1588.

DeMarco, T. and Lister, T. A. (1987). *Peopleware.* New York: Dorset.

Dunnette, M.D.(Ed)., (1976). *Handbook of Industrial and Organization Psychology.* Chicago: Rand-McNally.

Fox, J. M. (1982). *Software and Its Development.* Englewood Cliffs, NJ: Prentice-Hall.

Hastie, R. (1987). Experimental evidence on group accuracy. *In* G. Owen and B. Grofman (Eds), *Information Processing and Group Decision-Making.* Westport, CT: JAI Press, pp. 129-157.

Jones, C. (1986). *Programming Productivity.* New York: McGraw-Hill.

Kernaghan, J. A. and Cooke, R. A. (1986). The contribution of the group process to successful group planning in R&D settings. *IEEE Transactions on Engineering Management,* **33(3)**, 134-140.

Kidder, T. (1981). *The Soul of a New Machine.* Boston: Little, Brown.

Kling, R. and Scacchi, W. (1982). The web of computing: Computer technology as social organization. *Advances in Computers,* vol. 21. Reading, MA: Addison-Wesley, pp. 1-90.

Kraft, P. (1977). *Programmers and Managers: The Routinization of Computer Programming in the United States.* New York: Springer-Verlag.

Lemos, R. S. (1979). An implementation of structured walkthroughs in teaching Cobol programming. *Communications of the ACM,* **22(6)**, 335-340.

Lucas, H. C. (1984). Organization power and the information services department. *Communications of the ACM,* **27(1)**, 58-65.

Mantei, M. (1981). The effect of programming team structures on programming tasks. *Communications of the ACM,* **24(3)**, 106-113.

McGarry, F. E. (1982). What have we learned in the last six years? *Proceedings of the Seventh Annual Software Engineering Workshop* (SEL-82-007). Greenbelt, MD: NASA-GSFC.

Mills, H. D. (1971). *Chief Programmer Teams: Principles and Procedures.* Technical Report IBM-FSC 71-5108. Gaithersburg, MD: IBM Federal Systems Division.

Myers, G. J. (1978). A controlled experiment in program testing and code walkthroughs/inspections. *Communications of the ACM,* **21(9)**, 760-768.

Rich, C. and Waters, R. C. (1988). Automatic programming: Myths and prospects. *IEEE Computer,* **21(8)**, 40-51.

Rogers, E. M. and Kincaid, D. L. (1981). *Communication Networks: Toward a New Paradigm for Research.* New York: Free Press.

Scacchi, W. (1984). Managing software engineering projects: A social analysis. *IEEE Transactions on Software Engineering,* **10(1)**, 49-59.

Scott, R. F. and Simmons, D. B. (1975). Predicting programming group productivity: A communicatons model. *IEEE Transactions on Software Engineering,* **1(4)**, 411-414.

Sells, S. B. (1963). An interactionist looks at the environment. *American Psychologist,* **18 (11)**, 696-702.

Sells, S. B. (1966). Ecology and the science of psychology. *Multivariate Behavioral Research,* **1**, 131-144.

Shaw, M. E. (1971). *Group Dynamics: The Psychology of Small Group Behavior.* New York: McGraw-Hill.

Shneiderman, B. (1980). Group processes in programming. *Datamation,* **26(1)**, 138-141.

Srinivasan, A. and Kaiser, K. M. (1988). Relationships between selected organizational factors and systems development. *Communications of the ACM,* **30(6)**, 556-562.

Swanson, E. B. and Beath, C. M. (1988). The use of case study data in software management research. *Journal of Systems and Software,* **8(1)**, 63-71.

Thamhain, H. J. and Wilemon, D. L. (1987). Building high performance engineering project teams. *IEEE Transactions on Engineering Management,* **34(3)**, 130-137.

Tushman, M. L. (1977). Special boundary roles in the innovation process. *Administrative Science Quarterly,* **22(4)**, 587-605.

von Mayrhause, A. (1984). Selecting a software development team structure. *Journal of Capital Management,* **2(3)**, 207-225.

Vosburgh, J., Curtis, B., Wolverton, R., Albert, B., Malec, H., Hoben, S. and Liu, Y. (1984). Productivity factors and programming environments. *Proceedings of the Seventh International Conference on Software Engineering.* Washington, DC: IEEE Computer Society, pp. 143-152.

Walston, C. E. and Felix, C. P. (1977). A method of programming measurement and estimation. *IBM Systems Journal,* **16(1)**, 54-73.

Walz, D. (1988). A longitudinal study of group design of computer systems. Unpublished Doctoral Dissertation. Austin: Department of Management Science and Information Systems, The University of Texas.

Walz, D., Elam, D., Krasner, H. and Curtis, B. (1987). A methodology for studying software design teams: An investigation of conflict behaviors in the requirements definition phase. *In* G. Olsen, E. Soloway and S. B. Sheppard (Eds), *Empirical Studies of Programmers: Second Workshop.* Norwood, NJ: Ablex, pp. 83-99.

Weinberg, G. M. (1971). *The Psychology of Computer Programming.* New York: Van Nostrand Reinhold.

Chapter 4.2

Research and Practice: Software Design Methods and Tools

Barbara Kitchenham[1] and Roland Carn[2]

[1]*National Computing Centre, Oxford Road, Manchester, UK*
[2]*Reliability Consultants Limited, Fearnside, Little Park Farm Road, Segensworth, Fareham PO15 5SH, UK*

1 Introduction

This chapter will briefly review the current state of programming in commercial and engineering practice. It will attempt to summarize the work reviewed in the preceding chapters and to place it in perspective against the industrial scene. From this perspective some ways forward will be suggested.

The central message of the chapter will be that there is a need to relate cognitive mechanisms and human errors to errors committed during the software development process which are manifested as faults in the software.

We will start this chapter by reviewing current practice, the nature of programming, and the causes of human error. We will then describe some of the popular methods and tools for software design and specification, and highlight some of their advantages and disadvantages. We will conclude by discussing some of the challenges that study of experienced programmers offers to cognitive psychology, and the benefits that the software engineering community might expect from such studies.

Psychology of Programming
ISBN 0-12-350772-3

The tone of this chapter will be necessarily forward looking and speculative, in contrast to the soundly based work so far discussed. Little new work will be presented.

Most of the academic literature concentrates on the characteristics of novice programmers, and on how novices become experts. However, the characteristics of the professional programmer and of the expert programmer are of more interest to the commercial and economic community. Who, then, is a programmer and what is a program and what is programming?

Programming is practised at different levels by different people. Almost anyone can write a program and get it to work but there are dramatic individual differences between programmers. There are the novice programmers who have either never written a program or who have written only a few relatively trivial programs. Amateurs are often computer buffs for whom the technique is more important than the program or the application. Amateurs can be expert programmers and can sometimes write very high-quality systems. The casual programmer is professional in some, often technical, area but not a professional programmer. He is interested in the application; in using the computer to solve a professional problem. The class of professional programmers or software engineers is distinguished from the others in that the members of the class earn their living by building software systems. Generally, they have had formal training in programming and have a greater experience than the other groups.

However, it is not clear who the professional programmer is. In the commercial and financial industry, DP staff include programmers, analysts and designers. In the scientific and engineering industry staff are referred to as software engineers, software designers and programmers. It is recognized in these industries that there are differences in the skills, experiences and responsibilities associated with the job titles and this is reflected in the rewards attached to the job titles. The situation is further confused by the inconsistent use of the titles from organization to organization. So that a programmer in one organization simply translates detailed designs into code and tests the result for executability while in another organization the programmer is responsible for finding and implementing a computer solution to the user's problem.

The practising software engineer is responsible for the specification, design, coding, testing, implementation, documentation and maintenance of the software component of a system or equipment which contains software. The professional programmer as a practising engineer needs to understand the software development process. He will use this understanding to manage and control the errors that are committed during software production. He will then know what risks are associated with the use of his software. The users of the software take responsibility for it and are held accountable for their decisions.

Software does not exist in isolation from the hardware that it animates or from the environment with which it interacts. Software engineers work with professionals in other disciplines ranging from bankers and teachers to space scientists and military personnel.

The term software is a very broad term encompassing the executable machine code, the source code in a high-level language, the design documents, the test and management documents, and the user manuals for operating and maintaining the code. It refers, therefore, to the whole of the non-physical component of a computing system.

The definition of what is or is not a program is not always clear. In some situations, especially where the computing system is small, the term program is used to refer to the whole of the software. In other situations the term refers to a small subroutine of only a few lines of code. In practical terms a program could be defined as the smallest set of instructions that can be compiled and executed to perform a single coherent function. In a system of any size, it might be very difficult to identify such units when 'include' files are used and common service routines are shared.

The tasks required to create a program are not entirely clear. There is general agreement that there is some sort of thinking to be done to understand the user's requirement. Somehow one achieves a design and expresses it in a computer language. Finally one compiles, tests and tweeks it until the user can be persuaded to accept it, or not.

2 Current practice

Most commercial and industrial software systems are such that several different people are involved in their development. The development takes place in an organization which has socio-economic goals and structures. It is a social as well as a technical and commercial exercise. The engineering of software to meet the functional, performance and commercial goals of the organization is managed very much like other engineering activities.

The waterfall model described by Boehm (1981) is widely used as the basis for managing the development of software. This is a device that enables managers to do their management tasks. It is not based on any understanding of the process of software creation nor does it reflect the real activity of the people involved in the process (cf. Visser, 1988). As a consequence the history of software application is littered with failed, over-run and over-spent projects.

The creation of software remains in practice essentially a craft industry. Even in very large projects involving billions of pounds, and millions of lines of code where hundreds of software engineers are involved, design and programs remain the work of single individuals. Curtis (1989) has identified the 'super-designer' who is central to the success of large projects.

There have been several attempts at imposing a discipline on the process, which have been motivated by professionalism and by the need to manage the process. The structured methods have proved popular and successful partly due to the marketing zeal of their originators. Many of the structured methods are supported by graphics which have an intuitive appeal to humans whose primary modality is vision, particularly when several relationships need to be considered in parallel. Mathematical methods are currently being proposed because of their conciseness and the ability they offer of reasoning with the symbols of their arcane notation. However, it is not yet clear that any of these methods are closely related to the actual process of software creation.

In the pursuit of efficiency and effectiveness various tools have been developed and used to automate the repetitive and labour-intensive aspects of computing. Languages, compilers, environments and other specialized tools are in reasonably common use. They are valuable and are born of practical experience out of necessity. It is interesting to note that these tools are themselves software – thinking tools for thinking. More effective tools could be built if we understood the creation process and knew what tools were needed to support human weaknesses and magnify human strengths.

The often cited rate of technical change in software is all in the areas of new and more complex applications and the supporting hardware technology: faster processors, bigger, faster memories, more powerful computers. The organizational impact and the whole life cost of the software are now the most significant costs of computing. In spite of new developments, Cobol is still the most widely used language. The use of new methods presents too great a risk and too great a cost for the perceived return. In the scientific and engineering industries the situation is a little different. Here retraining is beginning to emerge as an issue. In anticipation of demand from the defense industry, some companies have invested heavily in ADA, VDM and Z.

Closer inspection of the software production process suggests that it is an engineering discipline like any other engineering discipline. It is not as mature as electrical or chemical engineering or even agriculture. Nonetheless, it is *not* an art form. Before software engineering can mature as an engineering discipline, practitioners need a better understanding of the process by which software is created in response to a demand and of the risks and errors which are associated with the process.

2.1 Nature of programming

The conventional view of the software creation process treats it as an engineering process that progresses from conception to realization through the stages of requirement definition, specification, design, construction (coding and unit testing), integration and operation. Conventional software engineering wisdom is that software is created in a top-down manner through progressive stages from the abstract requirement definition to the concrete realization of the final code. To this simple linear model is added the feedback and control necessary for management of the process. Thus each phase is verified to be in some sense a correct implementation of the previous phase, the specification is verified to meet the requirement, the design is verified to meet the specification and so on. Quality control also verifies that the further phase was reached by the agreed application of methods and procedures. Various stages of testing during the integration phase validate that the product meets the perceived need. Typically, validation and verification at each phase are associated with rework as faults and inadequacies are discovered.

Perhaps the most obvious characteristic of software creation is iteration. Iteration occurs at each level in the top-down model. It also occurs in an uncontrolled way in the verification-rework cycle and the testing-rework cycle. Our inability to predict and control these iterations is a major contributor to the chequered history of software applications. We need to understand the management and engineering tasks at each phase of the software development lifecycle in terms of the psychological activities and stages by which the software is created.

In practice, a computer system and the software that it contains is required to solve someone's information-processing problem. Thus a program is intended to achieve a specific goal, such as providing information for a decision or controlling some other mechanism, by manipulating data. The program is the set of instructions addressed to the computing engine, which is a perfect idiot, for performing the desired manipulations. Declarative descriptions of the data structures and the rules for applying the necessary manipulations are subservient to the procedural structure. The procedural structure is the product of decomposing the program goal into subgoals at progressively more detailed levels until the means to achieve the detailed goal can be expressed in a language that, after translation (compilation), can be interpreted by the computer. Until the advent of advanced architecture machines, multiprocessor systems and concurrent processing, the underlying computing engine was always assumed to be a Turing machine with a von Neuman architecture.

There are several problems associated with this view of programming. It fails to account for the order of magnitude differences in productivity between individual programmers. There is evidence, discussed by Visser and Hoc in Chapter 3.3, that the actual activity of programmers does not conform to the top-down model. There is no indication of how to control iterations. No account is given for the introduction of faults into the program. There is no way to determine whether a program is complete or correct because completeness and correctness are operationally undefined. We can only inspect designs and code to determine completeness and correctness. Although low-level errors can be detected by compilers and by executing the code serious errors of commission or omission in requirement, specification or design can often remain undetected and can be very expensive to correct. There is no way to determine which problems are hard and which are easy because we do not understand the mechanisms of program creation and cannot classify problems in terms of their solutions. For similar reasons there is no way to determine which design or program code is better than another. No account is given of the group dynamic and social factors that influence programming, discussed earlier by Curtis and Walz, in Chapter 4.1.

2.2 Cognitive models and software production

In order to create an effective piece of software a series of conceptual or cognitive models have to be created, co-ordinated and communicated to and among the members of the development team (and their managers). The nature, quality and integration of these models is critical to the eventual success of the software system.

The cognitive models illustrate and arise from the varying viewpoints that users and developers have of a software system:

The users' conceptual models of their problems/requirements determine a system's goals, and the way in which it will be used. A user's model must be communicated to and understood by the system designers/analysts. The programmers and designers need a conceptual model of the behaviour of the specific computing device or system on which the software will be executed. The programmers and designers also need a model of the virtual computing device that underlies the language in which the program is to be written.

These conceptual models can be related to the levels of abstraction known as the top-down development method and to the waterfall model of development. Users and designers create a set of requirements which a software system must provide. Designers/analysts create an information-processing model to solve the user's problem, which must be communicated to the programmer. Programmers create a textual model of the designers' information processing model, in the programming language. The textual representation of the system is then (automatically) compiled into machine code.

Only when a designer or a programmer has assimilated appropriate conceptual models of the programming task does he or she move from the ranks of novice to the ranks of expert software engineer. Thus, acquiring appropriate conceptual models underlies the conversion from novice to expert.

2.3 Cognitive models and human error

Human performance and human error result from the same cognitive activities. These activities operate at three levels:

(1) exercising a skill, where routine actions are performed effortlessly;

(2) organizing a collection of ideas and perceptions around a kernel concept, arrived at, perhaps, by analogy;

(3) solving problems, by setting goals and organizing activities into plans to achieve those goals.

Errors that occur when using a skill give rise to 'slips' when perceptual or motor patterns interfere with each other. Errors that occur at the insight level, give rise to incorrect identification of objectives and of key elements in the problem solution, and invalid assessments of the relationships among those elements. Errors at the problem-solving level result in incorrect goal structures and inappropriate or invalid plans. Errors relating to the last two levels are failures of knowledge and/or logic, and give rise to 'mistakes'.

Cognitive activities of each kind are needed to specify, design and program computers, and are the origin of the faults that are introduced into software systems during their production.

From this point of view a good development method is one which facilitates performance while minimizing errors. Good methods facilitate:

(1) problem understanding, specification, and elaboration;

(2) development of solutions (plans) that satisfy the problem requirements (goals) and satisfy complementary goals relating to different view points;

(3) controlled search of the solution space;

(4) recognition and control of incompleteness and inconsistency;

(5) movement among levels of abstraction/detail without loss of important information.

3 Software design

A software design is a plan for converting a specification into executable code, i.e. the means by which a problem description is turned into a problem solution.

Software design involves:

(1) Structural design which determines the components of software systems in terms of their required functionality and their links to one another.

(2) Algorithmic design which determines how each structural component delivers its required functionality. It is algorithmic design that is usually equated with programming in the way it has been described by Pair in Chapter 1.1.

Structural design may involve many levels of system decomposition before components are ready for algorithmic design (e.g. system-subsystem-module-procedure). It is an understanding of the nature of structural design that distinguishes an experienced program designer from a novice. (This is reflected in programmer's grading structures, where the ability to take over structural design responsibility distinguishes the system designer from the system programmer.)

All software systems are information-processing systems, they accept data as input, manipulate that data (often within the context of other previously input data), and provide data as output. From a technical viewpoint, design must, therefore, address three issues:

(1) definition of the data to be held by the system;

(2) definition of the process by which inputs are manipulated (i.e. the order in which system components are invoked);

(3) definition of the states that the system can assume and what transformations are permitted between states.

It is often the case that one issue dominates the other for a particular application (problem type). Data definition dominates many commercial systems. Process and/or state definition dominates many scientific, operating and embedded systems.

3.1 Requirements of design methods and tools

A design method needs to provide:

(1) a method and/or heuristic guidelines indicating how to create a design;

(2) a method for recording the design unambiguously (i.e. a notation);

(3) a method of, or procedures for, verifying that a design is correct (verification involves ensuring that the design fulfils the terms of the specification, is internally consistent and does not prevent or hinder further system development, i.e. coding and testing.)

Design tools support design methods, but more importantly support design as part of a software development process. Tools, therefore, have additional requirements to:

(1) support design configuration and change control;

(2) interface with specification, coding, and testing tools;

(3) make the design and the design process visible both for management and quality control.

3.2 Major design methods

Today, existing design methods split into two main types:

(1) mathematical methods (often referred to as 'formal methods');

(2) structured methods.

The mathematical methods are aimed at specification activities. They are intended for use in determining the system specification and component specification. The issue of structural design is not usually addressed explicitly.

They provide a mathematical specification notation and use the classic methods of mathematical proof to verify that a program is consistent with its specification. The process of design becomes that of 'reification'. This is a process of refining data structures and their associated operations from the abstract forms used in the specification to forms capable of being expressed in programming languages. For example, a specification might define a data item as a sequence but a programming language might only handle arrays. Reification ensures that the transformation of the data item from sequence to array is done in a way that can be proved to be correct.

In terms of the three requirements of a design method discussed above, the mathematical methods perform well with respect to providing an unambiguous notation and the possibility of verification. However, they do not offer many rules or heuristics for the *creation* of a specification.

There are many design approaches which come under the general heading 'structured'. They include:

(1) Entity-relationship modelling which considers the objects about which the system collects and maintains data. This approach is used for database design (Veryard, 1984), and is now being used extensively in the design of 'object management systems' needed as part of software engineering environments. [NB. An object management system is one that handles a variety of different object types, e.g. 'files', 'tools', etc., as well as the 'records' which a database system would handle.]

(2) Data-flow analysis and design, which at the analysis level, models the flow of data through a system and at the design level specifies the system structure in terms of the calling relationships among modules. Data-flow analysis is used to identify the functional and data components in a system and the order in which the functional components are invoked (DeMarco, 1978). Structured design methods associated with data-flow analysis refine the functional components into modules and define the calling relationships between modules (Yourden and Constantine, 1978).

(3) Data structure methods, which are based on analysing the structure of the input and output data. They are usually used to design individual functional components (Jackson, 1975), but have been extended to cover systems analysis (Jackson, 1983).

(4) Object-oriented design, which is based on identifying the objects of interest in a system and treating those objects as abstract data types with associated operations to interrogate or alter their state (Pressman, 1987). This approach is very closely related to the entity-relationship modelling approach.

Unlike the mathematical methods, the structural methods are particularly useful for the specification of system structure.

In terms of the criteria for design methods, the structural methods vary, but they are all weak on the issue of verification methods. The structured design approach of Yourden and Constantine (1978) offers a number of heuristics with which to assess a particular design based on the principles of coupling and cohesion (strength). The coupling principle advises designers to minimize the links among modules by restricting them, whenever possible, to direct calling links, and minimizing the use of common data structures. The cohesion principle suggests that modules should, whenever possible, perform a single, well-defined function.

All the methods, apart from object-oriented design, offer graphical notations that are intended to assist in the early process of design creation, but which are ambiguous and incomplete as representations of a design. The JSP notation for algorithm design (Jackson, 1975) is the least ambiguous and is supported by a design notation ('schematic logic') formal enough to drive automatic code generators.

In addition, the JSP method offers a very well-defined procedure for design creation and refinement. Unlike most design approaches JSP is teachable and repeatable – most programmers who use the method on the same problem will produce similar programs. This is not the case for any other methods, including the mathematical methods.

It is interesting to note that the strengths of the structural methods complement the weaknesses of the formal methods and vice versa. It is, therefore, surprising that there is no effort among software engineers to integrate the two approaches. This may be because historically (with the exception of Mascot) the structured methods have been developed by the commercial data-processing community and the formal methods have been developed by the scientific and real-time communities. It may also be a result of an emphasis by software tool builders on solving specific technical problems (e.g. imprecise specifications, or unreadable system descriptions) rather than investigating the underlying causes of the technical problems.

3.3 Tool support for the major design methods

Tools to support structured methods include:

(1) data dictionaries which identify each named item in a system;

(2) graphics systems;

(3) object-oriented languages with complex support environments, such as Smalltalk (Kay and Goldberg, 1976).

Tools to support the mathematical models include structured editors which ensure that specifications are syntactically correct, specification animators, specification refinement aids and programs that check (or even generate) proofs (Lindsay, 1988).

However, most of these tools are experimental and, with the exception of syntax checkers and animators, are not widely available to practitioners.

Work being undertaken to base tool design for formal methods support on detailed task analysis is a particularly interesting development (Masterson *et al.*, 1988a,b). Masterson and his colleagues adopted the knowledge acquisition techniques used in the development of knowledge-based systems (KBS) and the ideas of user-centred design (Monk, 1985) to gain an understanding of how an expert creates a specification.

In common with most software development activities, writing a specification in a formal language involves an iterative cycle of exploring the problem space and then representing it in the language. Masterson *et al.* investigated the way experts explored the initially ill-defined problem space, and the way the experts refined the problem space by their attempts to express it clearly. They used the information to guide the design of tools to assist formal specification, and have developed tools to support VDM (Jones, 1986), and Z and are in the process of designing tools to support Lotos.

They describe the process of knowledge acquisition related to creating Lotos specifications in some detail (Masterson *et al.*, 1988a). Lotos (ISO/DIS8807, 1987) is a language that allows the specification of concurrency by use of a process algebra derived from CCS (Milner, 1980) and CSP (Hoare, 1985).

The interviews took the form of asking Lotos experts:

(1) about their experience with the language, in particular the way in which they structured the task of writing a specification;

(2) to look at a Lotos specification and work out what it did, providing a commentary of what they were looking at and why;

(3) what features they would like to see in tools.

The interviews indicated the relative importance of features of the specification during the creation of the specification (i.e. understand the processes before the data typing), and the ordering of activities (i.e. identify and elaborate the component processes, before assembling the top-level behavioural description; only consider data typing after the construction of most of the process algebra). In addition, the style of specification that achieved maximum clarity was identified (i.e. no more than two levels of embedding; structuring and specification around the 'normal' case, using simple synchronizations).

Masterson *et al.* conclude that tools to support Lotos should include: a means of representing the specification based on graphics and cross-referencing among gates, syntactic aids, specification reorganization aids, data-typing aids, library aids, an animator, and a temporal logic aid.

Tools that integrate all the various aspects of software development are called SEEs (software engineering environments), and have been subject to extensive research efforts in Europe and the USA during the past five years.

Such tools place stringent requirements on their designers to provide a consistent, usable interface for a variety of users (i.e. not just software engineers, but software project managers and quality managers as well). In addition, SEEs maintain permanent records of existing designs, and so could be designed to assist the use of past experience which Visser and Hoc noted as a usual expert design strategy (Chapter 3.3).

Most software engineering environments concentrate on providing support for technical processes. However, the Programmer's Apprentice is an important exception (Rich and Waters, 1988). The long-term goal of the Programmer's Apprentice is to develop a theory of how expert programmers analyse, synthesize, modify, explain, specify, verify and document programs. It is intended to apply artificial intelligence techniques to automate the programming process. The work centres on two basic principles, the assistant approach and inspection methods (cliches); and two main technical advances, the Plan calculus and a hybrid reasoning system (Cake).

The assistant approach is based on the idea of assisting programmers rather than replacing them. It suggests that each programmer should be provided with a support team in the form of an intelligent computer program that can take over routine programming tasks. It assumes a substantial amount of communication between the programmer and the assistant based on a substantial body of shared knowledge of programming techniques. The Programmer's Apprentice is viewed as an *agent* in the software development process not a tool.

Inspection methods are the ways that expert programmers construct programs from basic components (i.e. plans or cliches). Inspection methods are based on the assumption that experts construct programs by inspection rather than reasoning from first principles. In analysis by inspection, properties of a program are deduced by recognizing occurrences of cliches and referring to their properties. In synthesis by inspection, program construction is achieved by recognizing cliches in specifications and selecting an appropriate implementation cliche.

The Plan calculus is a formal representation for programming cliches. It is part of the hybrid knowledge reasoning system called Cake. Cake allows users to manipulate frames and cliches and to use logical and mathematical reasoning.

Although it is clear that plans are not sufficient to account for expert performance (see Chapter 4.2), the principles underlying the Programmers' Apprentice appear to offer many facilities that actively support expert behaviour. Petrie and Winder (1988) observed that experts develop their own pseudocode to represent partial solutions in different notations and at different levels of detail. Visser (1988) observed that experts do not follow a strict top-down approach. These aspects of expert behaviour are supported by the flexibility of representation handled by Cake, as well as the ability to store and reuse expertise using libraries of cliches.

4 Cognitive issues

The cognitive tasks involved in software design are mainly those concerned with problem solving but also include learning issues.

4.1 Problem solving

Problem-solving tasks raise a number of issues which might be best investigated from the viewpoint of cognitive psychology:

(1) Selecting the best design method for a particular software application.

(2) Identifying design notations that are an integral part of the design creation. This involves investigating the extent to which the search for a problem solution is assisted by the notations used to describe the problem and its solution. From a software engineering viewpoint, currently only the JSP techniques match a problem definition with a method of problem solution.

(3) Identifying design notations that enhance a programmer's ability to recognize a correct problem solution. This involves investigating the extent to, and ease with, which a design can be verified for completeness and correctness.

4.2 Learning

A designer must obviously learn the basic design notations and approaches, but also needs to determine the most appropriate approach for a particular problem class.

A major controversy surrounding the mathematical methods is whether they can be learnt and used by more than a minority of software engineers. In this debate, the issue of declarative as opposed to procedural approaches raised by Pair (Chapter 1.1) is also relevant.

The mathematical (and indeed the object-oriented) methods are declarative. However, the current generation of software designers in industry have been brought up with the procedural programming languages. This means that the difficulties that software engineering researchers experience in gaining acceptance for new approaches may not be blind ignorance or laziness as is often asserted, but a genuine cognitive difficulty caused by the interaction of an unfamiliar notation and an unfamiliar underlying approach to design representation.

This presents a challenge to the cognitive psychology community to extend their work on skill acquisition of novices to retraining skilled staff. The concept of representation and processing systems developed by Hoc would seem to be a fruitful approach to this problem.

Learning tasks include not only learning the basic method, but also tasks that arise from the fact that in large software developments different people are involved in different stages of the development process. Thus, a designer may need to read and understand a specification produced by another person before producing a design, and a design may, in turn, be passed on to other people for coding and testing. Thus, a notation must be not only be unambiguous, it must also be readable and understandable.

The mathematical notations usually force precision at the expense of redundancy (although the Z method provides specifications that include both a mathematical and natural language section). The structured methods use graphical notations and natural language, and so provide redundancy at the risk of ambiguity.

In addition, a software designer is often faced with the task of incorporating new functions into an existing system, which involves 'learning' a system to the extent that new functionality can be safely incorporated.

The learning task is often made more difficult by the fact of variability among experts noted by Visser and Hoc. Understanding a design developed by another designer often involves understanding a solution that you yourself would never have envisaged, given the particular problem statement. This implies that software engineers should aim to develop design methods and tools that reduce variability.

5 Discussion

A major problem for software designers is identifying the appropriate design approach for a particular application area. We need better criteria for evaluating our methods, and those criteria must include factors such as how well the method supports the cognitive activities involved in software design, both from the viewpoint of initial design creation and from the viewpoint of design readability.

Another major problem is that of the quality of software products. Software development is a human-intensive process which is therefore subject to human errors. We need to ensure that our methods and the tools that we use to support those methods support human problem-solving activities while minimizing the opportunity for 'mistakes' and 'slips'.

Therefore, when developing tools to support methods, it is necessary to consider the cognitive tasks that must be supported as well as the technical tasks. Masterson *et al.* (1988a,b) are developing intelligent tools to support formal methods, which were designed after a detailed analysis of the tasks performed by an expert. There would seem to be a wide application for this approach to tool development.

The software industry as a whole needs to overcome its retraining problems. The new techniques being developed in the universities and the research laboratories are very different from those used by software practitioners. The knowledge of novice skill acquisition that cognitive psychologists have developed must be extended to the area of expert retraining. Unless serious efforts are made to retrain existing staff, the adoption of new methods, particularly the mathematical methods, will be severely delayed.

References

Boehm, B. W. (1981). *Software Engineering Economics.* Englewood Cliffs: Prentice-Hall.

Curtis, B. W. (1989). Video Presentation. Psychology of Programming workshop, Warwick University, Warwick, UK.

DeMarco, T. (1978). *Structured Analysis and System Specification.* Englewood Cliffs: Prentice-Hall.

Hayes, I. J. (Ed.) (1987). *Specification Case Studies.* Englewood Cliffs: Prentice-Hall.

Jackson, M. A. (1975). *Principles of Program Design.* New York: Academic Press.

Jackson, M. A. (1983). *System Development.* Englewood Cliffs: Prentice-Hall.

Jones, C. B. (1986). *Systematic Software Development Using VDM.* Englewood Cliffs: Prentice-Hall.

Kay, A. and Goldberg, A. (1976). *Smalltalk 72. Instruction Manual.* Palo Alto: Xerox Research Centre.

Lindsay, P. A. (1988). A survey of mechanical support for formal reasoning. *Software Engineering Journal,* **3**, 1.

Hoare, C. A. R. (1985). *Communicating Sequential Processes.* Englewood Cliffs: Prentice-Hall.

ISO/DIS8807 (1987). Information processing systems – open systems interconnection – LOTOS – a formal description technique based on temporal ordering of observational behaviour.

Masterson, J. J, Ishaq, K. P. and Hockley, A. T. (1988a). An approach to providing support tools for Formal specification. *FORTE '88*, University of Stirling.

Masterson, J. J., Ishaq, K., Patel, S., Norris, M. T. and Orr, R. A. (1988b). Intelligent tools for formal specifications. *Proceedings in Software Engineering*, **88.**

Milner, A. J. (1980). *A Calculus of Communicating Systems*. Berlin: Springer-Verlag.

Monk, A. (Ed.) (1985). *Fundamentals of Human-Computer Interaction*. New York: Academic Press.

Petrie, M. and Winder, R. (1988). Issues governing the suitability of programming languages to programming tasks. *In* D.M. Jones and R.L. Winder (Eds), *People and Computers*. Cambridge: Cambridge University Press.

Pressman, R. S. (1987). *Software Engineering. A Practitioner's Approach*, 2nd edn. New York: McGraw-Hill.

Rich, C. and Waters, R. C. (1988). The Programmer's Apprentice: a research overview. *Computer*, **21**, 11.

Veryard, R. (1984). *Pragmatic Data Analysis*. Oxford: Blackwell Scientific.

Visser, W. (1988). Giving up a hierarchical plan in a design activity. INRIA Rapports de Recherche.

Yourden E. and Constantine, L. (1978). *Structured Design*. Yourden Press.

Index